Numbers Smaller than 20

Elementary school math! Can we learn in a more interesting way that is more meaningful to us?

Many students think math is boring and difficult. It is not that unusual for us to meet people who gave up on math. Given this reality, this book was made with a teacher's enthusiasm in teaching students elementary math in a more interesting and meaningful way.

This book contains at least three characteristics. First, based on Korean math and education procedures, materials are compared to Chinese, American, and Japanese math curriculum and analyzed. The vital elements of elementary math were extracted. Therefore, you can learn general math content through this book series. Second, we developed the manuscript by considering how students acquire math content and apply the steps and methods to do so, even examining their behavior with math. Third, considering students' interests and curiosity those factors extracted are reflected in an interesting fantasy story.

Researchers and writers of this manuscript took a meaningful journey during its preparation. We anticipate that all students who are reading this book will go on a journey with Ryan and Aris to learn math and realize that math is fun and meaningful in our world.

Korea National University of Education (Math Department) Professor
Bang Jung Sook

I will become a revolutionist who will shake the stereotype of **studying being boring!**

While I was in school, I didn't know the reason why I had to learn math. Memorizing the formulas and solving questions repeatedly, I loathed math at my young age. After becoming an adult, I was surprised that all mathematical principals are blended into our life. In reality, math has a close relationship with us, yet our children do not know such facts and don't see the connection. Therefore, they lose interest and even get stressed on solving questions and the scores they get. Eventually, they lose confidence in math.

This led to the forming of a team of math teachers, education specialists, and professional writers to make *Ryan's Math Adventure*, offering students a way to gain confidence and enjoy math.

We analyzed the education procedures of Korea, China, America, and Japan and made a conceptual and global curriculum. And through the deployment of a fantasy adventure, we solve mathematical questions and connect the abstract ideas of math and our reality. Children forget that they are learning math and enjoy this comic book style, naturally understanding the principles of math within. They complete study goals by using a workbook to establish concepts and solve interesting questions. Also, through a fantasy world with various types of beings and cultures and a European style of illustration, without violence, that we consider necessary for children to have a positive mindset and behavior toward math and a multi-language system, I hope our children grow up to become global human resources for our bright future.

WeDu Communications CEO
Lee Kyu Ha

Composition & Features

Integrated Global Education course

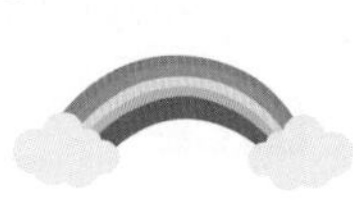

Through comparative analysis of math curriculum of Korea, China, Japan, and America, the Integrated Global Education Course is designed for a new math "edutainment" book. With contents intended to reflect world trends and to provide the objective rationale, the Integrated Global Education Course can target a broad range of readers.

Mathematical Contents, Elements, and Composition

❶ Based on the math learning achievement standards presented in the curriculum of China, Japan, and America, we have developed concrete mathematics contents which elementary students should learn.

❷ By analyzing the other three countries' curriculum based on Korea's math learning achievement standards and textbook contents, we have identified the most commonly emphasized math contents, which are reflected in the book.

❸ Other details contained in the other three countries' curriculum but not in Korea's are also identified; noteworthy differences are captured in the book for a truly in-depth aggregate composition.

Goals of Integrated Global Education Course

❶ In the course of observing, analyzing, organizing, and expressing various phenomena around us, learners will develop abilities to discover mathematical concepts, functions, principles, and laws inherent in those phenomena and to understand the correlations between them.

❷ The capability to mathematically think, express, and communicate will be fostered; learners will be able to solve various issues derived from different phenomena in creative and reasonable ways.

❸ Learners will be able to understand values of mathematics; they will look upon math as a joy; and a good-natured personality will be nurtured.

Most children regard mathematics as a boring subject where memorization and calculation are heavily involved. Unlike their expectations, *Ryan's Math Adventure* provides fun math learning through storytelling techniques which help learners in learning math concepts and principles while reading comics. Learners will be able to adopt a flexible, creative, mathematical way of thinking while exploring a fantasy land.

1

By closely correlating fun fantasy stories and math concepts, this book empowers children to understand mathematical principles in a fun and natural way.

2

Through daily life events, children will learn how deeply math is related to our everyday life.

3

The main characters of the book deal with not only the direct mathematic area but also mathematical competence emphasized in elementary school mathematics curriculum including problem solving, reasoning, and mathematical communication, and more. Therefore, as each character grows in math capability and good manners, learners will achieve the same growth.

Comprehensive exercises will be provided to see how well children understood the contents.

An explanation matching to each curriculum is provided to help learners gain a full understanding of math concepts.

Extended exercises will be provided for children to have an in-depth understanding and to let them get used to various types of math practice.

Introduction of characters

Ryan
(10 years old)

A normal student
from our world.
Loves fantasy novels, games,
cartoons,
and movies, but hates math.
Usable item: Yo-Yo

Aris
(7 years old)

A curious princess
from Tanare Palace
that lives in this
fantasy world.
King Adel,
who is her father,
gives her the *Star of Wisdom* as a gift.
Usable item: Magical Wand

Phillip
(10 years old)

A noble man
from the fantasy world.
Excellent with numbers
and operations.
Like his peers,
Phillip is excellent with magic
and swordsmanship.
Usable item: Sword

Numi
(10 years old)

A member of the Elf Clan,
loves adventures and
is very active. Has very good
knowledge of
diagrams and uses
diagram magic.
Usable item: Bow

Pabel
(10 years old)

A member of the Dwarf Clan which had the knowledge of measurement but, after the magical curse of Pesia, lost all knowledge of math.

Usable item: Axe, Hammer

Gilly
(13 years old)

A member of the Floa Clan who can transform into a tree. Knows basic math and learned magic.

Usable item: Asian Guitar

Walter
(33 years old)

Royal general of Tanare Palace. Excellent with math and magic. Good with machines and builds a robot for Aris.

Namute

A smart robot created by Walter to protect Aris. But after interacting with Ryan, Namute becomes a weird robot.

Special Skill: Transform into a motorcycle

Villains of this book

Pesia

A person full of desire to become the only king and to rule all. With the *Staff of Chaos* in his possession, he erases nearly all knowledge of math from the world. He threatens the journey of Ryan and Aris, looking to get Aris's necklace, the *Star of Wisdom*, at any cost.

Usable Item: *Staff of Chaos*

Sirocco

A faithful servant to Pesia, he comes from the same town as Walter. However, Sirocco is always number 2, due to Walter, so he has great anger and jealousy toward him.

Dagan

He is close to Lord Dior of Onix but his true identity is as servant of Pesia. He works as an informant and passes the secret of the *Book of Light* to Pesia.

Nurimas

The only blood relative of Pesia - a nephew. Since he was young, he has been raised by Pesia so that now he is cold and decisive, and obeys Pesia no matter what.

Past story

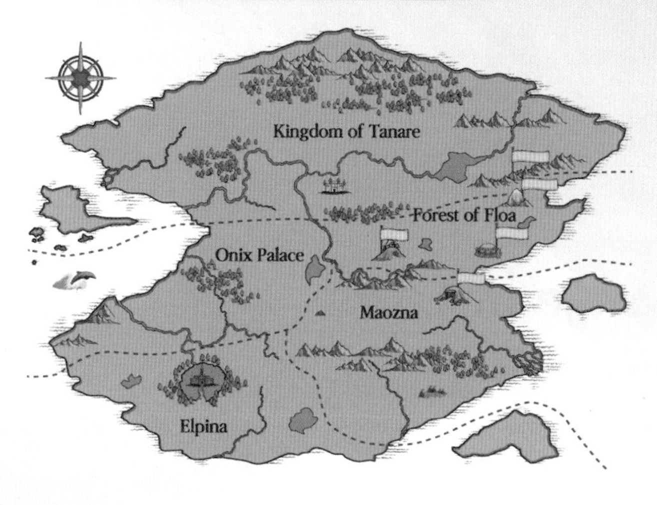

Ryan is someone from our world who hates math.
He teleports to a fantasy world with a book he found at the museum. Meanwhile, the peaceful Kingdom of Tanare has been overtaken by a villain named Pesia. Pesia uses his Staff of Chaos to erase all knowledge of math from the world. A world without math is a world in great disorder. Our heroes Walter and Aris barely make it out of the castle and away from Pesia. Ryan meets Walter and Aris and together they start a journey. They find the Book of Light, just like the old prophecy said. Ryan, Aris, and Walter are heading to Walter's hometown, Revna. They meet a girl named Gilly, part of the Floa Clan, who can transform herself into a tree. And they also meet Pabel of the Dwarf Clan, who was lost and was being chased by a swarm of bees. And Numi, part of the Elf Clan, who was hungry and looking for some food. Everything is quiet and peaceful at Walter's home. Then someone knocks on the door...

contents

1. A change in town

① Counting, Writing, and Reading Numbers from 1 to 5 ··· 12
② Ordering from 1 to 5 ··· 24
③ Understanding the Idea of 1 More and 1 Less (1) ··· 29
④ Understanding, Writing, and Reading 0 ··· 34
⑤ Counting, Writing, and Reading Numbers from 6 to 9 ··· 37

2. Workers at the robot factory

⑥ Understanding the Order of Numbers up to 9 ··· 44
⑦ Comparing Numbers ··· 49
⑧ Understanding the Idea of 1 More and 1 Less (2) ··· 52
⑨ Comparing Numbers up to 9 ··· 58

3. Stone of Life

⑩ Different Ways of Grouping the Numbers 2, 3, 4, 5 ··· 70
⑪ Different Ways of Grouping the Numbers 6, 7, 8, 9 ··· 80
⑫ Understanding, Reading, and Writing 10 ··· 82
⑬ Grouping and Gathering the Number 10 ··· 85
⑭ Counting, Writing, and Reading Numbers from 11 to 19 ··· 89

4. Birth of Namute

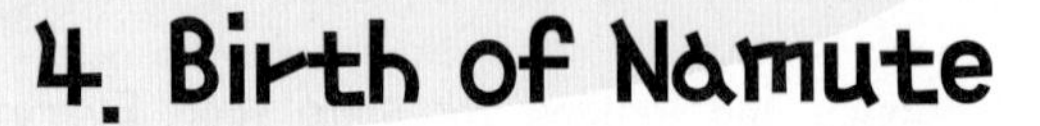

⑮ Grouping and Gathering Numbers from 11 to 19 ··· 100
⑯ Understanding Numbers up to 19 ··· 102
⑰ Comparing Numbers up to 19 ··· 108
⑱ Understanding Even and Odd Numbers ··· 116

1. A change in town

Who is it?
Knock
Knock
Knock

Captain Walter!
It's me, Phillip!
Phillip?

What brings you here?

My grandfather sent me.
Go to Walter's house.
So you've come all the way from Onix?
No, I was in Revna.
I see.
Where's Aris?
I think she's still sleep...
Hey, Phillip!

Why're you here?
Did you miss me?
Yeah... k-kind of...

Are you going to play with me today?
I can't stay long...

You're leaving already?
I'll give you a gift instead!

A gift? What is it?
Here you are...

I'm late
for school!

Who's he?
I need to
get to school!

?
Ooooops!
Bump!

Ack!
Ryan!
Are you
okay?
Thump!
What's his
problem?

What class are you in?
What is he talking about?

Ryan, stop! What's wrong?
Oh... Aris...

That's right! This isn't...
Were you dreaming?

Who're you?!
What?

Say hello, Ryan.
This is Phillip, the grandson of Lord Dior of Onix.
Humph. Your name's Phillip, huh?

Look at this!
Phillip gave me a gift.
What's inside?

I was about to open it.
It's nothing special.

Wow!
They're cherries!
There are tons of cherries at the supermarket...

"Supermarket"? What's that?
?
Nevermind.

I have to go now.
Leaving so soon?
All right.

You'll come again, won't you?
S-sure I will.

Next time we'll go to Phillip for a visit.
Really?

Please tell Lord Dior that we're safe here.
I will.

Also tell him that we will meet with him soon.
I'll tell him.

We should go into town ourselves.
Why?

There are a few things I have to buy.
Yipee! Going to town!

By the way, how many cherries are in there?
I bet there are a lot!

It's a small pouch.
I guess so.

There are this many cherries in here.
Huh? What is this?

I'm confused. Try writing it as a number.
Hmm...

You don't know your numbers, do you?
Numbers? What are those?

I'll show you.

1 One
Okay! You write "1," and read it as "One."
2 Two
This is "2," and you read it as "Two."
3 Three
Now you write "3," and read it as "Three."
This is "4," and you read it as "Four."
4 Four
5 Five
This is "5," and you read it as "Five."
If you can remember these, you'll be counting soon!
Wow... This is amazing!

One
Two
Three
Four
Five
Now, do you want to count how many cherries there are?
There are five cherries altogether.

What are you two doing?
Sir Walter! I learned about numbers.
I taught them to her!

Where did you learn
math, Ryan?
At school,
of course!

Was it a
magic school?
Just a regular old
school.

I see. Let's start heading
out to the village.
How exciting!

I'll leave these
cherries here.
I'll leave
my bag here, too.

This is the town of Revna.

Wow! So this is a town!
This place is giving me a bad feeling.

What're those kids doing?
Hey, mister, I'm first!
No, I'm first!
I'm first!!
Huh?

That old man is
still here.
Who is he?

He hands out magical balloons
to the children of the town.
Magical balloons?

I want one,
too!
A-Aris!

I want one too,
mister!
Who're you?
Quit
shoving!

Haha, if you kids keep this up, I won't be able to hand out any balloons.
I told you guys! I'm first!
Nope! Me first!
I wanted one first!
No one's gonna get anything if you don't stand in line.
What is 'line'?
Line?

Let's try lining up in the order you got here.
Who was here first?
I was.
And Aris came last, right?
Yes, that's right.

If you line up in order like this, you can all get a balloon.

What does 'in order' mean?
Remember I showed you how to count to five?

Ah, so the boy at the front of the line is number 1. The boys after him are 2, 3, 4.
That makes me number 5, doesn't it?

When you use numbers to show order, you say:
first,
second,
third,
fourth,
and fifth.

Awesome! My balloon is a dragon!
Mine is a tree! My favorite!

Wow!
My balloon is shaped like a flower!
It's perfect for you.

We should buy some food now.
Like what?

There's a bakery over there.
Bakery
Blech! We're just getting bread...What about some sandwiches?

What are sandwiches?
They're meat and veggies in between two slices of bread.

That sounds fake!
They're real! We sell them in restaurants and sandwich shops!

You say the weirdest things...
You're driving me crazy!

Ryan, can you buy some bread at the bakery? I'll get some clothes in the store next door.
Fine!

I bet the bread here stinks.
I'll go with Ryan!
Wait for me back here after you get the bread.

I hope they have all sorts of tasty bread!

Excuse me, ma'am, I'd like to buy some bread.
There are many to choose from. Go ahead and pick what you want.

Huh? There's barely any bread here!
We have some red bean buns, cream buns, and chocolate buns.

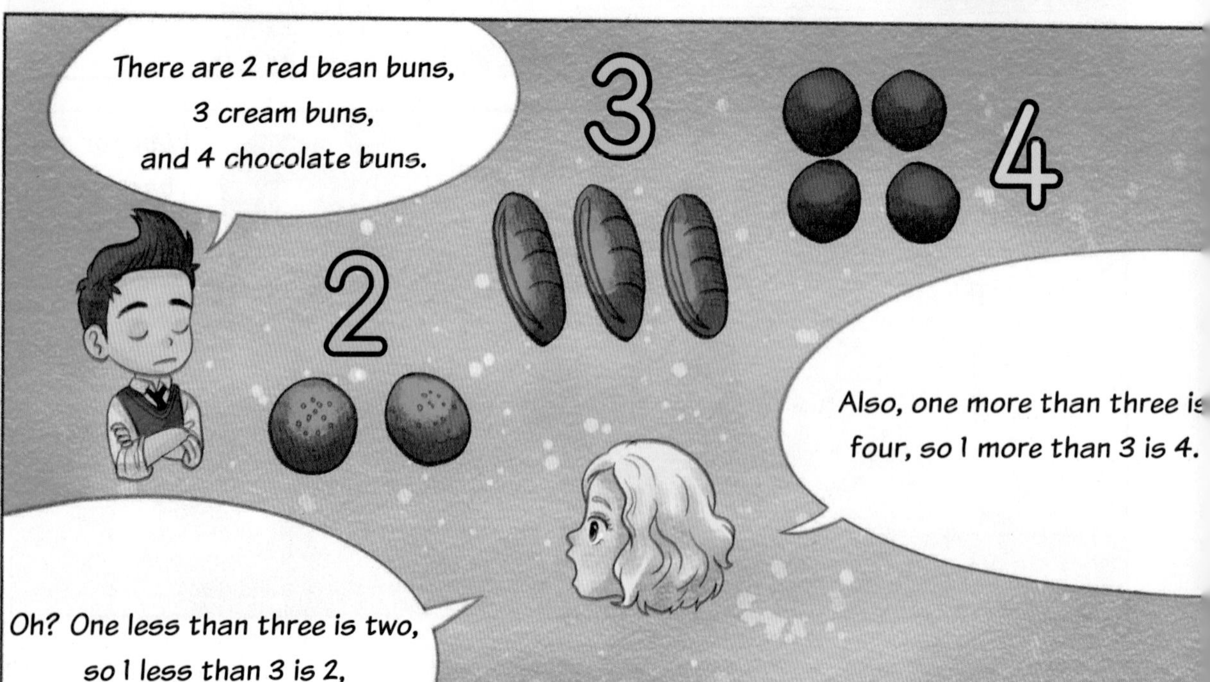
There are 2 red bean buns, 3 cream buns, and 4 chocolate buns.
3
4
2
Also, one more than three is four, so 1 more than 3 is 4.
Oh? One less than three is two, so 1 less than 3 is 2, right?

That's right!
One less than 2 is 1.
One more than 2 is 3.
1
2
3
4
5
Aha!
1 less than 4 is 3.
1 more than 4 is 5.

You pick up math very fast, Aris!
Huh? Have I been learning math?

Learning numbers up to 5 and
lining items up in order are both math.
Wow! I learned math!!

Are you two going to buy some bread?
We'll take everything you have.

How much is it altogether?
You don't know how much your bread costs?
Hmm... Well... How much is it...?
I guess not. Why don't you just take them?

We can't just take them. This should cover it.
Th-thank you.
That was weird. How could she not know the price?
At least we got the bread!

What's the matter, Ryan?
The woman at the bakery couldn't tell us how much the bread cost.
Bakery

The man at the clothing shop couldn't tell me the price too...
Has this town always been like this?

Revna was never like this.
Something must've happened.

I need to go see someone about this. Follow me.
Okay.

We're here.
Let's go in.
Wow! There are so many strange things here!

What are you looking at, Ryan?
There's a store across the street selling milk.

We should get some milk to go with the bread we just bought.
You go ahead then! Meet us back here.
I'll go with him.

Welcome!
Do you have any milk?
Ding
Ding

There is plenty of milk here!
Take as much as you want.
Ack! What're you talking about?!

Here you are!
You can put as many bottles as you want in here.
There are only 2 bottles here!
That's not plenty!

Let's just get them both.

Hmm?
What are you staring at?

There is no more milk left.
Is there a number for 'nothing'?
There is! You write it as "0" and read it as "Zero."
2
Two
1
One
0
Zero

Hehe. I understand now.
How much for the milk?

I was hoping you could tell me!
You don't know how much your milk is?

I have no idea...
Here! Just take this!

I understand.
I won't tell anyone where the princess is.
Creak
This town doesn't make any sense!

The owner there didn't know the price either?
Yeah! And she said there was plenty of milk. She had only 2 bottles!

The townspeople have been acting strangely.
Since when?

Since this morning...
This morning?

They seem to have forgotten all about math.
Math?
Have you forgotten as well?

Of course not.
Strange indeed.

I know math, too! Ryan taught me!
You must be Aris! Nice to meet you.

Would you like to try some? I made them myself.
What are they?

They're cookies.
Wow!
My favorite!
I like cookies,
too!

I have more.
Take these.
Cool!
Thank you!

We have more to
talk about.
Yes, why don't you
eat your cookies
over there?
Yes, sir!

By the way, who is that
boy?
I think he's lost.
He's far from
his home.

How many cookies do you have?
You have your own!

Well, I was gonna give you some if I had more.
Really?

I should count mine then.
How many? A lot?

Hmm...
What's wrong? Too many?

What do you call the number that's 1 more than 5?
Oh, that's right! I taught you only the numbers up to 5, didn't I?

6
Six
You write one more than 5 as "6," and you read it as "Six."
How about one more than 6?
You write one more than 6 as "7," and you read it as "Seven."
7
Seven
You write one more than 7 as "8," and you read it as "Eight."
8
Eight
You write one more than 8 as "9," and you read it as "Nine."
9
Nine
Aha! I got it.

I have 9 cookies.
I also have 9 cookies.

Then you and
I both have 9 cookies,
right?
You got it!
We have 9 cookies each.

Sir Walter,
I learned more math!
It's time
to go home.

You did? Ryan taught
you again?
Yes.

Ryan, can you use
magic by any chance?
Of course
I can't.

2. Workers at the robot factory

Good bye.
Get home safely!
Farewell!
Bye!

Who is that man, anyway?
When he was young, he used to work at the magic school.

He knows magic?
Yes, indeed.

I just wanna eat.
I want to learn magic, too.
Why does he remember math?

Wh-what's
going on here?
What's all that noise?
I hurt my arm!
I hurt my back!!

Everyone, calm down!
The doctor needs to
see me first!
Nope!
Me first!
Everything hurts
right now!
I should be
seen first!

All those men want to see
the doctor first.
I think I've seen
this before.

Right! With the magic balloon...
We lined up in order then, didn't we?

I think I should go and help.
Let me go with you.
But I'm starving...

Good afternoon, doctor!
Oh! Is that you, Walter?

What's with all the noise?
I'm not quite sure.

I think I may be able to help.
Could you? That would be so kind of you.

Everyone needs to stand in a line.
Line? What's a line?
I don't know what that is! I just want to see the doctor first!
I'm first! Me!

The doctor can't help anyone if you keep shoving!
That's right! You have to make a line starting from the first person who came!

I was here first.
I arrived right after.
I guess I'm here...
I came last.
If you line up like this...
You'll all see the doctor in no time!

I'll send them in one at a time.
Thank you! I can finally help the patients.

You two go help the doctor.
Sure!
Hmph!
These men seem to be workers from the robot factory...

I'll start handing out
the medicine. Take this number,
and wait over there.
Thank you so much.

Why are you
smiling?
This is my first time
in a hospital!
It's so amazing!

I'm tired of this
place already!
You're not badly hurt.
I'll be okay?

Take this number,
and wait over there.
2

Now that I've finished treating everyone, I need to give medicine to all of these people.
You'll hand them out by following the numbers, right?

Yes, that's right.
I'll give them their medicine. Get some rest.

You're number 1, right? Here's your pill.
Thank you.

Next is number 2. Here you go.
Thank you so much.

That's strange...
What is?
That man was the second person in line.
Doesn't that mean Sir Walter should give him 2 pills?
Being the second person in line means...
he's the second person to receive medicine.

Aha! I understand now.
How much longer are we gonna be here...Phew...

Great work, doctor!
Thank you. With your help, I was able to finish quickly.

The patients look like workers from the robot factory. Is that right?
You are correct.

Is it normal for this many workers to get hurt?
No, this is unusual.

I think something must've happened at the factory.
I think I should go to the factory and see for myself.

Oh no! I forgot!
What's wrong?

I forgot to give the patients on the second floor their medicine.
I guess you have to see them now.

I need to go to the second floor.
Is there anything that I can do?
I'll help you, too!

What do you mean? You just want to take a look around.
No, I really will help!

This is the patients' room.
All these men are sick?

I'll bring the medicine.
I'll bring it!

Sure.
The medicine is over there on that table.
Okay!

There are 6 patlents, so I should get 6 pills, right?

No, this patient went to the bathroom.
Don't forget to get one for him, too.

1 more than 6 is 7...

I need to bring 7, right?
Yes, that is correct.

She's quite smart. Are these your children, Walter?
Wrong!
N-no, they're not.

They all take the same medicine?
They're magical pills. They can cure any sickness.

Here is another patients' room.
I see.

All of these women need medicine, too?
That's right.

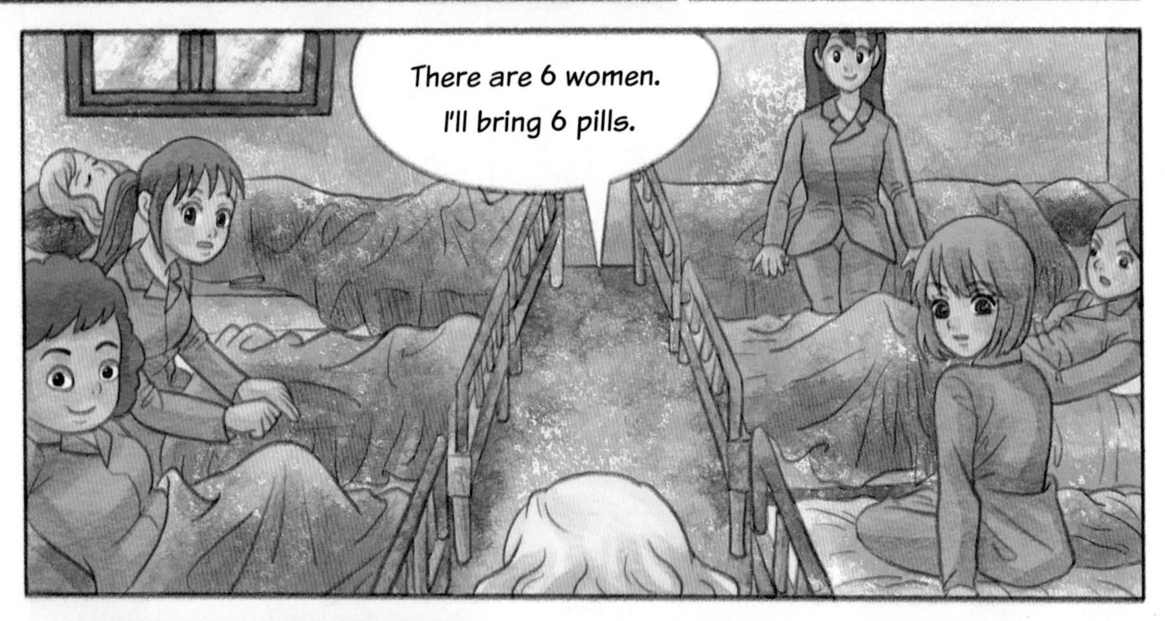
There are 6 women. I'll bring 6 pills.

Dear child, I'm all better and will leave soon. I don't need any medicine.
??

1 less than 6.

I need to bring 5 then.
Looks like you're having a blast.

You were once a wizard, weren't you?
That is ancient history.

Have you forgotten everything about math?
Not a bit. I remember it all.

Good bye!
Have a safe trip home.

Are we going to the robot factory now?
Yes, I should see if anything has happened there.
I'm still hungry...

The robot factory is not far from here. Hold on a little longer.
Whatever! I'm just gonna eat these cookies!
I'll have some, too!

Something fishy is going on in this town.

Was it different before?
Revna was a very peaceful and happy town.

It's that Pesia guy's fault!
!!!
Huh? How do you know about Pesia?

I saw him in my dream. That bad man...
Ryan, I need to have a word with you.

have never seen you at the palace.
Who are you, really?
I'm just a student.

Did Pesia send you here to spy on us?
No! Never! I don't like that man either!

Fine...Don't mention
Pesia again to Aris.
Not a word.
I think she needs to
know though.

You have to
promise me!
Okay...I promise!

Humph! She'll figure it
out sooner or later...

Ryan, Sir Walter, what're
you two doing?
N-nothing.
I'm... just... hungry...

Waaaaaah!
Huh?
There is a child crying over there.

Why are you crying?
Sob! Sob!

All my parents do is fight with each other. They don't even give me food when I ask for it!
You're crying because you're hungry?
I know how you feel!

Take these cookies.
Cookies?
What?

If you give him your cookies,
what're you going to eat?
I can have some bread
when we get home.

Here, you can have
mine instead.
You said you were
hungry.

Let's give him the bag
with the most cookies!
Fine with me.

I have 4 left.
I have 7 left.

I have more cookies than you.
Correct! I have fewer cookies than you.
7
4
7 is a greater number than 4 and
4 is a smaller number than 7.

I will give him my bag.
Here! You can have my cookies, too.

He'll need more than cookies.

Here! Some bread and milk, too!

You said you were hungry!!
He's younger than I am. I'll survive.

You're so kind!
All right already!
He's not as bad as I thought.

Look over there!
What is it?

They're so pretty!

I've always wanted one of these...
Buy it then!

Which should I get? Which looks pretty on me? Huh? Huh?
I don't care.

What's in those boxes?
3
6
Hairpins!

Why are you selling them in boxes?
Because they're so small. It's better to sell them in boxes.

I want some hairpins.
Just choose something already.

There are two boxes. Which one should I get?
They have the same hairpins inside.

Do you need a lot?
No, not that many.

Then why don't you get this one?
3
How do you know without opening them?

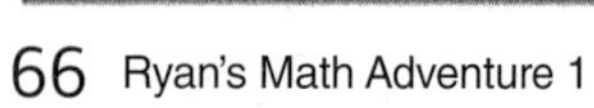

The 3 on this box means that there are three hairpins inside.
The 6 on this box means that there are six hairpins inside.
3
6
Three is less than six.
I see! And 3 is less than 6, right?

I'll buy this one!
3
Can we go now?
3
How much is this, mister?

It's free.
Really?

You don't know the price either?
What do you mean, "I don't know the price"? I'm giving this to her because she is such a pretty young girl.

Thanks so much, mister!
Suspicious...

Do you know magic by any chance?
I went to the magic school when I was a kid.
They are so pretty!!

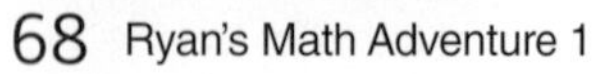

3. Stone of Life

Are we there yet?
We're almost there.

Look over there! A robot!

What is wrong with that robot?
It's acting a little weird...

We need to get to the factory fast.

There are more robots.

None of these robots are acting normal.
Something must be wrong with them.

I think the robots are all broken.
I don't think so. They're probably just poorly made.

Huge problem!
What should we do now?
I'm not sure...

Mr. Samson!
Oh! It's you, Walter!

What in the world is going on?
I'm not sure myself.

It seems like all the robots are broken.
It was a huge mess!

Are these your kids, Walter?
I'm just taking care of them for now.

There aren't
y problems with the robots.
Then why are they like this?

The problem is the workers.
What do you mean?

The workers don't remember the work order.
They can't tell apart the shape of the gears!
They can't even count how many parts are needed.

Has anything like this happened before?
No, today is the first time that this has ever happened.

I believe that they have all lost their knowledge of math.
It seems that way to me, too.

None of you can remember the work order?
Not me!
I can remember!
We're fine for some reason.

Can you all do magic?
Yes, we can.

Anyway, I sent those workers who cannot work home to rest.
The ones that got hurt are at the hospital.
Yes. I saw them in town.

We can't leave the robots like that.
I agree.

What're you going to do?
All of these robots have a Stone of Life in them.

If we take the stone out, they will stop moving.
Oh! I see!
How amazing!

I'll help you take out the stones.
That would be a great help, Walter!

I'm not sure if we'll be of any use, but...
We'll help you, too.

What does a Stone of Life look like?
It looks like this.

Then I'll try to take them out, too!
Ryan, wait for me!

Now on to
the next robot!

Where's the
Stone of Life?

There are 4 robots.
I'll take on 3, and you take on 1, okay?
Why don't we just do 2 each?

What are they doing?
I don't know...

Mister, need me to help?
Sure.

You take care of the robots here.
Yes! You do them all.
What??

When you put 2 and 2 together, it becomes 4.

Aha! I guess this is a Stone of Life!
You're right! It looks just like the one the man at the factory showed us.

I can't find the stone in this robot...
It's so hard to find when the robots are moving!

I should just look for another robot.
Hey Ryan, look over there!

What's going on?
Let's find out.

I'll deal with 5 robots, and you do 3, okay?
I think it would be better if we each handle 4.
Are you two arguing again?

Are you finished already?
Yep, all taken care of.

Then take care of these robots, too!
They're so silly!
Argh!

This time, when we put 4
and 4 together, we got 8.

I have no choice. I'll have to
take the Stones of Life out of
these robots.
But...
Ryan...

These robots are
a bit scary.
You think...?

What are they doing?!
Swing~
Crash!
Aaaah!
Boing

Whew! I wasn't expecting that!
There are more robots now!

We gotta find those stones!
Wait, Ryan!!

I can count up to 9,
but what is 1 more than 9?
1 more than 9 is written
as "10," and read as "Ten."

We have 10 fingers,
see?
I'll count my fingers
one by one.
One, Two, Three, Four,
Five, Six, Seven,
Eight, Nine, Ten.
One, Two, Three, Four, Five,
Six, Seven, Eight, Nine, Ten,
right?
10 Ten
You can group 10
ones together
like this!
I get it!

Here is the Stone of Life for this robot.

I have to take it out of the next robot now.
Ryan, let me try.

I'll stop the robots with my magic.
What? Magic?

Freeze!
Huh? What is it doing?
I think you should take a break from the magic.

Wow, that was
hard work.
We haven't taken all
the stones out yet.

I don't know where the
other robots are.
We've collected 10 Stones
of Life.

Let's take these
to Sir Walter.
That's what
I was thinking.

I'll help!
I'll carry 7,
and you take 3.
7
3
How about 5 for
you and 5 for me?
5
5
No,
I can carry more.
4
6
Then 6 for you
and 4 for me, okay?
Deal!
Let's do that!

You're seriously
a fast learner,
you know that?
Huh? You mean I just
learned more math?

Let me give you a problem to solve.
Sure!

I wrote down the numbers from 1 to 9 here.
I know. Why did you do that?
1 2 3 4 5 6 7 8 9
Can you choose which numbers make 10 when put together?
Ah! 10 is 1 more than 9. 9 and 1 make 10.
That's right!
And 10 is 2 more than 8. 8 and 2 make 10.
9 1
Oh, I see!
8 2

Math is
so much fun!
Give me your stones.
I can carry more.

You also have the
box of hairpins.
I'm fine.

It doesn't look like
anyone else is back.
Are we the first
ones to arrive?

I wonder what the inside of
the factory looks like.
Let's check
it out.

This is amazing!
Come on! All factories look like this!

So this is what a factory looks like! This is incredible!
I guess you haven't been to a lot of places, have you?

Ryan! Look over there!
Where are you going?

This robot looks cool!
I don't think this robot is finished.

I'll finish it for them!
Aris... what're you planning to do?

I'll get it started with my magic!
Not that magic of yours again!

Get moving, robot!
Don't!

Pow!
What's happening?
Hey!

It looks fine to me.
Did I use the wrong spell?

Something has to be broken...
Should I try again?

Crash!
Creek!
Crash!
Hum...
I knew it!

I think I heard something.
Is someone in there?
Oh no...
I guess Sir Walter is back.

He'll be furious!
Should we just hide these pieces?

What are you two doing over there?
There you are...
Us?

What happened to the robot that was hanging there?
I'm... sorry...

Wha...
What did you do?
I tried to... use my magic to...

Haha, that's okay! I was going to rebuild that robot anyway.
Why did you try to use your magic?

I wanted to play with the robot.
Oh? You wanted your own robot?

Well... We actually need a robot.
I can give you some materials to make your own robot.
Really?

Our factory can't make any more new robots at the moment.
As long as I have the parts, I can build one at home.

Hehe! Yipee!
We have robots back home, too!

What are these?
These are the Stones of Life we took out from the robots.

You have so many!
How do you count numbers that are bigger than 10?

1 group of ten and 1 one,
1 group of ten and 2 ones,
1 group of ten and 3 ones...
Hey!
What are you doing?

I have no idea how to count higher than 10...so...

How do you say 1 group of ten and 2 ones as a number?
10 2
2 10
10 2? Or 2 10?

That's not how you do it!
How then?

If you have 1 group of ten and 2 ones,
You write 1 to show the ten first and 2 for the ones after.
12 Twelve
You write "12," and read it as "Twelve."

Write	12
Read	Twelve

Write	14
Read	Fourteen

Write	15
Read	Fifteen

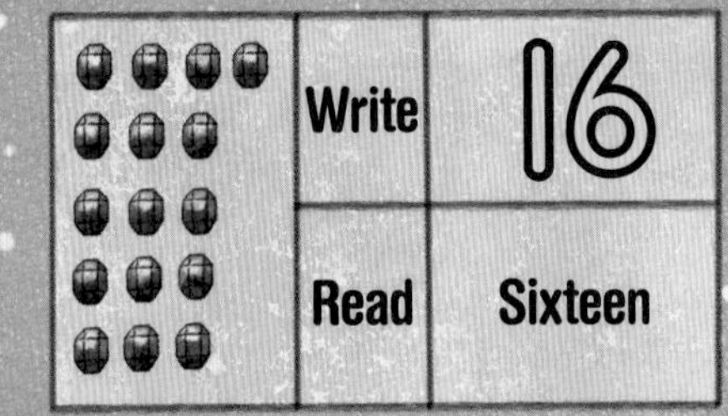

Write	17
Read	Seventeen

Write	18
Read	Eighteen

Write	19
Read	Nineteen

This is so easy
and fun!
Easy?

Yes, I think you're
a good teacher!
Then count to 13 using
the Stones of Life!

Okay!
I'll give it a try!

Tada!
Oh!
You got it!

4. Birth of Namute

Where's Sir Walter?
I'm not sure.
He went to the warehouse.

What for?
He said he was going to pick out a robot to take home.

Hey, kids, can you put 12 Stones of Life in this box here?
No problem!

I'll do it!
Huh? What're you up to?

I need to give you this many, right?

Where's the warehouse anyway?
Why do you ask?

We want to check it out.
Why don't you follow me then?

Is it far?
No, it's close.

There are so many robots here.
Wow!

Sir Walter! What are you doing?
I'm picking out a robot.

Can I choose it?
Sure, why not. Pick one.

11
12
13
14
15
Number 12 is so big!
14 is too small.

13
13 is perfect!
Looks a lot like a trouble maker.

Mr. Samson, we'll take this one.
Go ahead.

Now, we need to get the parts to finish the robot.
We'll go with you.

What are all these?
Those are parts to make a robot.

What are we taking?
Well, we need to make arms and legs.

But why do we need a robot?
For you, Aris!

For me?
To do chores at home and to be your friend.
I can be her friend.

I think these will do.
Can we make the robot now?

We need to finish making him at home.
Can you really make a robot at home?

The small building beside our house is a workroom. We can finish making the robot there.
Let's hurry up and go home then!

Let's pack everything up in this cart.

Oh no!
I forgot the number of the robot.
We can find it again.

What was the number?
12
This big one is 12.

And this small one is 14.
14
15
So the number for the robot that you picked is?

12
Aha! This one!! 13.
13
14

Let's take
e robot and
go now.
Okay!

All finished, Walter?
I just need a few more parts, Mr. Samson.

Hold on a minute. I'll get you what you need.
Thank you so much.

This is too much!
Let's put these over there.

What took you so long?
This one robot... was so far away...
I need some water...

I need a break.
I pulled a muscle.
There are so many Stones of Life!

Which box has more Stones of Life?

Well, I can't tell just by looking at them...

Let's take them out and count.
Let's group them in tens.

I'll count the stones in this box.
Then I'll do this one.

In each box, there is 1 group of ten.
When you compare 4 and 7, 7 is bigger.
The box I counted had more Stones of Life.
It's easier if you compare numbers like this!

Are you going somewhere?
I'm going to get some parts.
I'm going with you!
Me, too!!!

Wow! This place looks so cool!
It's just ok, if you ask me.

What're those?
16
19
We write down our daily tasks on those papers.

16
19
What are these numbers for?
They tell us how many kinds of robots we need to make.

What does 16 and 19 mean?
16
16 means 16 field worker robots. 19 means 19 chore robots.

With 16 and 19,
1 represents the tens.
Right?

16
Field Work
Robots
19
Chore
Robots

Yes! And 6 and 9
represent the ones.

6 is smaller than 9,
so 16 is smaller than 19.

16
Field Work
Robots
19
Chore
Robots

Correct!
They make more
chore robots!

16 is smaller
than 19.
19 is greater than 16.
Right!

You can show comparison
between two numbers
like this: 16<19.
16
Field Work
Robots
<
19
Chore
Robots
Ahh! I got it!

We can't make any robots today anyway.
Yes, we should go now.

Whoa! What is this place?
We store robot parts here.

All these are for robots?
Yes, indeed.

There are so many different kinds.
Weird looking, too!

What are these numbers here for?
HEADS 12
BODIES 14
LIMBS 8
Those are how many parts we need to make our robots.
HEADS 12
BODIES 14
LIMBS 8
It says 12 heads, 14 bodies, and 8 limbs.

How do you compare three numbers to see how big they are?
You can pair two numbers and compare them first.
Then I will compar
the size of 12 and
14 first.
BODIES 14

For 12 and 14, 2 is smaller than 4.
That means 12 is smaller than 14.

Then what about 14 and 8?
HEADS 12
BODIES 14
LIMBS 8

8 is greater than 4, so then is 8 bigger than 14?
Then you can say that 14 is greater than 8.
14 is 1 ten and 4 ones while 8 is just 8 ones.

14 is the greatest number.
14 is greater than 12,
and 12 is greater than 8.
14 is the greatest number.
8 is the least number.

And 8 is the least, right?

You two really like math!
I just started learning!
I'm her teacher!

Aha! I should try teaching math to my workers!

Do you need anything else?
I think... I have everything I need.

Can I take a few Stones of Life?
Take as many as you want.

Children, let's go home now.
Ugh, I finally get to eat!

That's right! You guys haven't eaten this whole time!
I'm going to faint...
Hehe! I'm feeling hungry, too!

How do they make a Stone of Life, Sir Walter?
You don't make them. You can get them from monsters or in mines.

You use them only to make robots?
Not really.
They have many different uses, such as lighting up the dark.

Ah! So it's like energy?
'Energy'? What's that?

Can I have one of the Stones of Life?
What are you going to use it for?
I want to put it in my yo-yo.
Yo-yo?
You'll find out when we get home.
Fine.

Phew! I feel so much better!
I can't eat another bite.

I'm so full, I can't move.
Your tummy looks so funny!

Can we have something that's not bread next time?
Revna is not what it used to be. You saw how hard it was for us to even get that bread.

I want some spaghetti with meatballs...
What's that? Is it food?

I want to go home... Sigh...
Do you not remember the way?

I don't know if I can ever find my way back.
Is your home really far?
Hmm...

How did you come to the forest?
I was holding that weird book, and then I stepped into the Fantasy Experience booth.

You're so strange, Ryan! I never have any idea what you're talking about!
Errr...

It's like you're from another world! Hehe!
Ouch! That hurts!

I'll be in
my workroom.
I want to go
with you.
Me, too!
Me, too!

The workroom
is dangerous.
Be careful.
I'll just look
around!

You brought these
boxes from the
factory, right?
What's in
them?

Wow!
These are for
making robots?

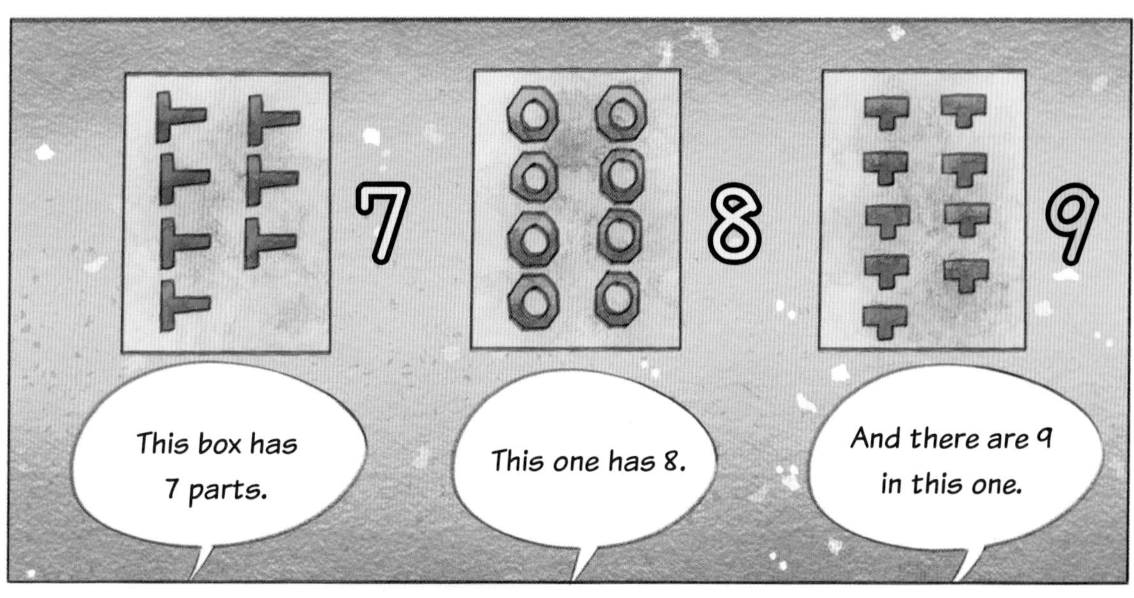
7
8
9
This box has 7 parts.
This one has 8.
And there are 9 in this one.

Want me to teach you some more math?
Really? Yes!

When you can make pairs like this, it's called an "even number"!
When you can't make a pair like this, it's called an "odd number"!

Do you still have a lot to do?
Yawn
I think just a bit more.

Why isn't this going in?!
You're busy, too?

What is that?
It's my yo-yo!

Is it a toy?
Don't make me laugh! People used to hunt with this in the old days!

Then do you use it to hunt?
No... I just play with it.

So... it's a toy...
Grrr!

Can you come over here, Aris?
Why?

I'm done! We need to pick a name for our robot.

You're all finished!

Why don't you name it?
Can I?

What should I call it?
What would be good?
I need a pretty name...

Namute! How about Namute?
As you wish!
Tada!! I'm done, too!!

The robot will be complete once I put in the Stone of Life.
Hurry and put it in!
Let's take you for a test spin!

I put the Stone of Life inside. It'll glow every time I throw it!

Okay! Here we go!
Yes! Hurry!

Time to yo-yo with style!

I'm putting in the stone now.
AAAHHHHH!! It's not stopping!!

What in the world!
Aaaah!!!
Smack!

Ryan!!
Ryan!
Zzzap!

Ryan,
are you hurt?
Ow...
Are you all right,
Ryan?

I'm fine!
Thanks for asking!
!!
??

It stung a bit,
but I'm okay.
Namute is
moving.

What are you staring at?
First time seeing a robot?

Nice weather today!
I should go for a walk!
...

Ryan,
are you okay?
Yeah, it felt like my entire body was tingling for a sec, but I'm fine now.

What happened?
This yo-yo just flew farther than I expected.

It would be best if you were more careful from now on.
Okay.

Why don't you go inside and get some rest?
No,
I'll just cool off for a bit.

Then I'll head in first.
Come on, Ryan.
Just a sec...

There is something wrong with this yo-yo.

There's no way I threw it hard enough for it to go as far as it did!
How could it go that far when the string is so short?

What about Namute?
Namute can find his way back home. There's no need to worry.
What is wrong with this?
How far can this go?
It has a mind of its own!

Poof Bustle Bustle
Hmm?
Is there someone
in the forest?

Is it a monster or
a wild animal?

Workbook
1

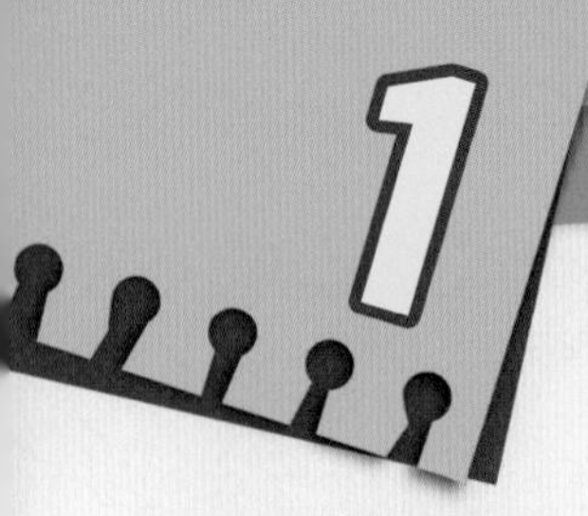

Numbers Smaller than 20

Counting, Writing, and Reading Numbers from 1 to 5

Comics for Learning Math

Phillip arrived at Sir Walter's house on an errand from his grandfather. Since Phillip could not stay long, he gave Aris a pouch of cherries as a gift.

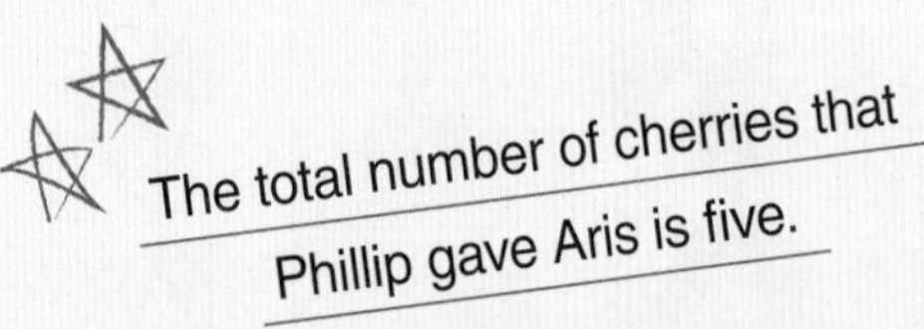

Review

You count and write numbers as 1, 2, 3, 4, 5, and you read them as "one, two, three, four, five."

		Write		Say
	/	1	1	One
	//	2	2	Two
	///	3	3	Three
	////	4	4	Four
	/////	5	5	Five

Warm Up 01 Choose the <u>incorrect</u> picture.

a. → /

b. → //

c. → ////

d. → ////

e. → /////

01-1 **Which pair of spoken and written numbers matches the fruit?**

a.

Two – 1

b.

Two – 2

c.

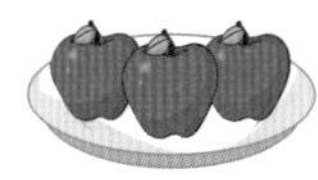

Three – 4

d.

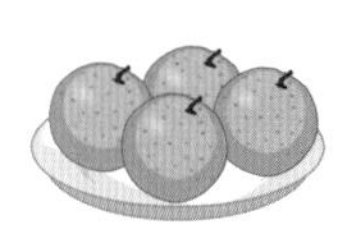

Four – 3

e.

Five – 4

01-2 **Look at the picture. Write the correct number on the lines.**

(1) There are ____________ glue sticks.

(2) There are ____________ pencils.

(3) There are ____________ pairs of scissors.

(4) There are ____________ sheets of colored paper.

Numbers Smaller than 20

Ordering from 1 to 5

Comics for Learning Math

The town of Revna is in commotion because the children all want a magical balloon from the old man. Even Aris wants one! Can Aris and Ryan figure out an orderly way to hand out the balloons?

Review

Order is expressed as first, second, third, fourth, fifth.

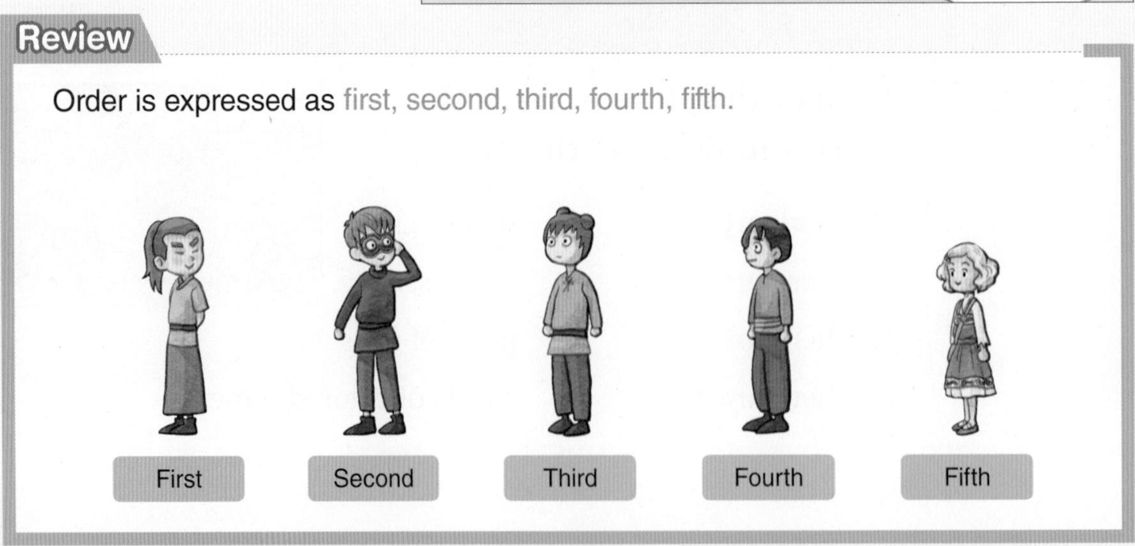

Warm Up 02 What is Alice's position?

a. First b. Second c. Third d. Fifth

02-1 **Choose the pair that correctly matches the number with its ordinal.**

a. One – Second
b. Two – First
c. Three – Fourth
d. Four – Fourth

02-2 **Look at the picture. Write the correct name on the lines.**

(1) ______________ will wash his/her hands first.

(2) ______________ is third in line.

Numbers Smaller than 20

Understanding the Idea of 1 More and 1 Less (1)

Comics for Learning Math

Ryan and Aris were hungry, so they went to the bakery to buy some bread. However, the bakery was nearly sold out of bread.

In the bakery, there were 3 cream buns. The number of sweet bean buns was one smaller than 3, which is 2. The number of chocolate buns was 1 more than 3, which is 4. That's what was left for Ryan and Aris to buy.

Review

Let's look more closely at counting up and counting down.

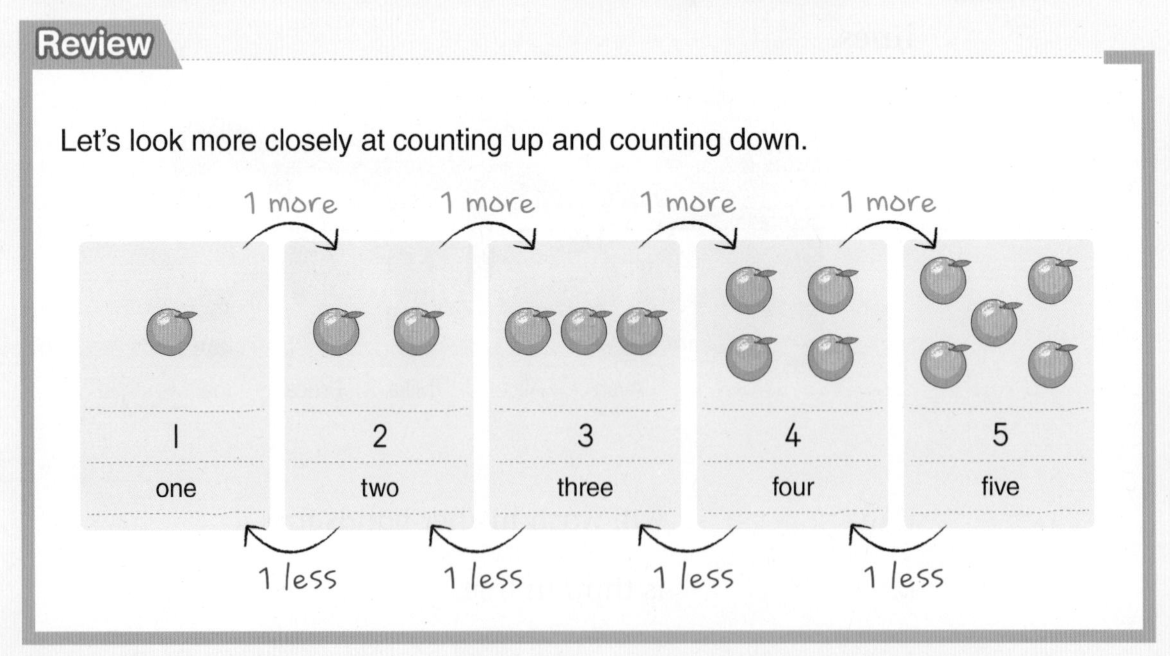

Warm Up 03 Mark ○ below the picture that has one more apple than the Figure 1.

( ) ( ) ( )

03-1 Draw ○ below the picture that has one more tree than the Figure 2, and draw △ below the picture that has one less than the Figure 2.

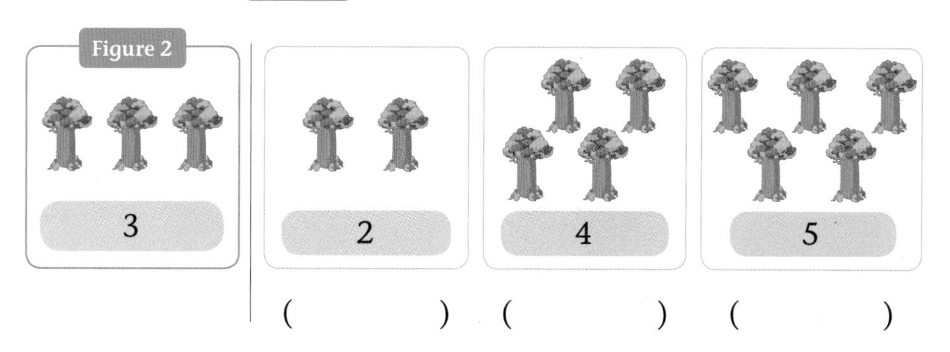

( ) ( ) ( )

03-2 Draw ○ below the picture that has one more than the Figure 3, and draw △ below the picture that has one less than the Figure 3.

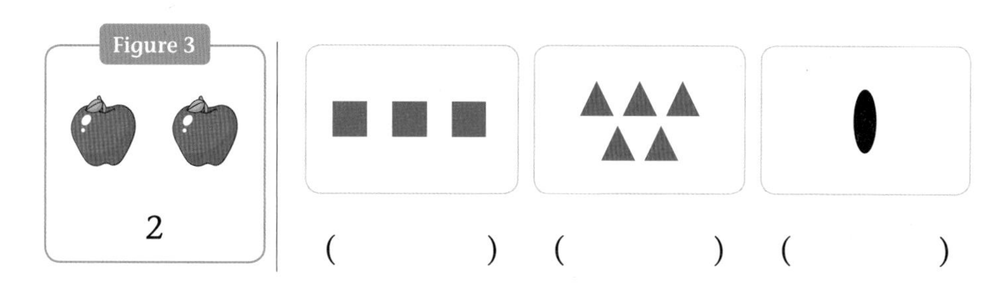

( ) ( ) ( )

Numbers Smaller than 20

Understanding, Writing, and Reading 0

Comics for Learning Math

Ryan and Aris went to another shop to get some milk for the bread they just bought. Ryan wants to buy all the milk in the shop.

If Ryan buys all the milk, there will be no milk left in the shop.

Review

If there isn't anything to count, we write 0 and read it as "zero."

 0  zero

Warm Up 04 **Read the story. Write the correct written and spoken number in the blank.**

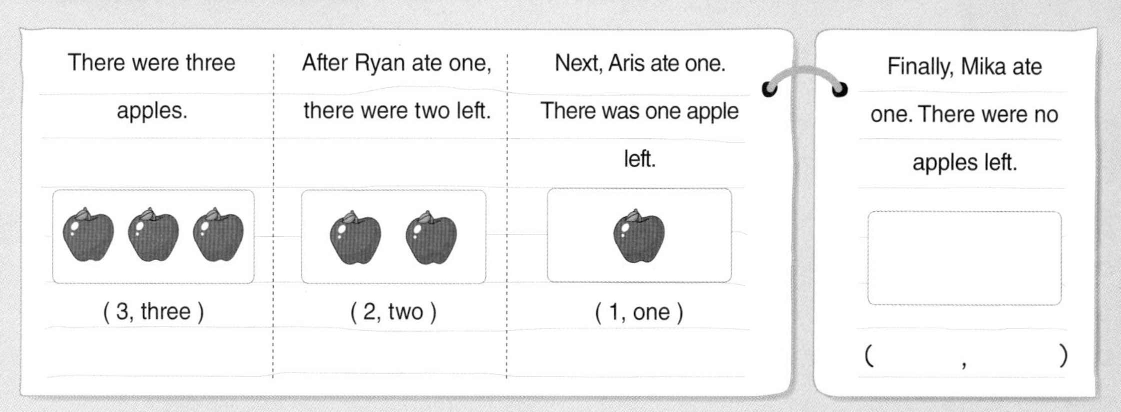

04-1 **Look at the milk counters below. Fill in the blank with the next number in the pattern.**

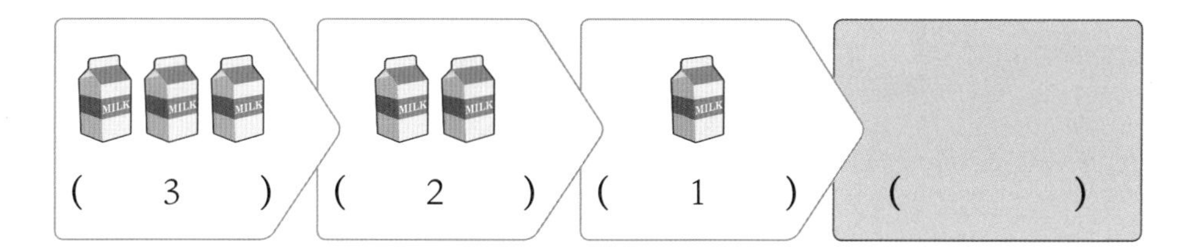

04-2 **Write the correct number in the blank.**

(1) The number which is one smaller than 1 is ______.

(2) The number which is one bigger than ______ is 1.

Numbers Smaller than 20

Counting, Writing, and Reading Numbers from 6 to 9

Comics for Learning Math

Ryan wants to count the number of cookies that the man gave him and then give some to Aris. Aris also wants to count the number of cookies she has.

The number of cookies Ryan and Aris each have is nine.

Review

You count 6, 7, 8, 9 and read them as "six, seven, eight, and nine."

		Write		Say
	//////	6	6	Six
	///////	7	7	Seven
	////////	8	8	Eight
	/////////	9	9	Nine

Warm Up 05 Count the fans. Write the number of fans on the line.

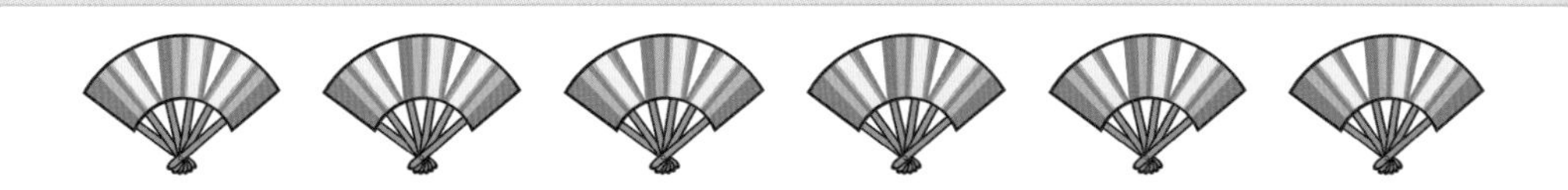

05-1 Draw a line connecting the stars to the matching number.

(1) ☆☆☆☆	•	• 8, Eight
(2) ☆☆☆☆☆☆	•	• 4, Four
(3) ☆☆☆☆☆☆☆	•	• 7, Seven
(4) ☆☆☆☆☆☆☆☆	•	• 9, Nine
(5) ☆☆☆☆☆☆☆☆☆	•	• 6, Six

05-2 Write the name of the person who said a true statement on the line.

Ryan: There are 6 ♡, and we read it as "seven."
Numi: There are 6 ♡, and 6 is one bigger than 5.
Aris: There are 7 ♡, and 7 is one bigger than 7.

Numbers Smaller than 20

Understanding the Order of Numbers up to 9

Comics for Learning Math

A loud group of men have gathered outside the doctor's office in Revna. They all want to be treated first and are fighting with each other.

Ryan and Aris are telling the men to line up in order. That way they can all be treated.

Review

We express ordinal numbers with first, second, third, fourth, fifth, sixth, seventh, eighth, and ninth.

Warm Up 06 **These are ordinal numbers. Write the missing numbers in the ☐.**

2nd - 3rd - ☐ - 5th - 6th - 7th - ☐

06-1 **Choose the ordinals that are in the correct order.**

a. fourth – fifth – seventh – sixth – eighth
b. fourth – fifth – sixth – seventh – eighth
c. fifth – fourth – sixth – seventh – eighth
d. fifth – fourth – seventh – sixth – eighth

06-2 **Look at the picture below. Write the correct ordinal number in the blank.**

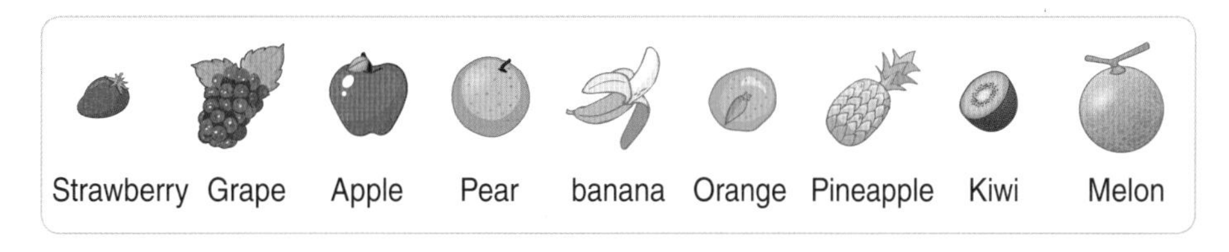

(1) The orange is ______ from the left.

(2) The kiwi is ______ from the left.

Numbers Smaller than 20

Comparing Numbers

Comics for Learning Math

Ryan and Aris are going to help the good doctor. They are trying to hand out medicine to each patient according to the patient's number. However, Aris thinks something is wrong.

The number written on the paper is that person's place in the order. If a man has a 2 written on his paper, that means he will be the second person to receive medicine.

Review

Compare the numeral and ordinal numbers.

Numerals	1	2	3	4	5	6	7	8	9
Numbers	One	Two	Three	Four	Five	Six	Seven	Eight	Nine
Ordinal numbers	First	Second	Third	Fourth	Fifth	Sixth	Seventh	Eighth	Ninth

Warm Up 07 **Look at the picture below. Who is correct? Write their name on the line.**

Aris: There are five strawberries in all.
Ryan: There are fifth strawberries in all.

____

07-1 **Color the apples according to the numeral and ordinal numbers.**

07-2 **Fill in the blank with the correct answer.**

(1) There are ______________ cars in total.

(2) The orange car is parked in the ______________ spot from the sign.

(3) The color of the seventh car from the sign is ______________.

Numbers Smaller than 20

Understanding the Idea of 1 More and 1 Less (2)

Comics for Learning Math

The doctor wants to give out medicine to the patients on the second floor. Aris wants to help again, but one of the patients has gone to the bathroom.

Aris needs to bring medicine for 6 male patients and 1 patient who went to the bathroom. 1 greater than 6 is 7. Of the 6 female patients, one of them is going to be discharged, so she needs to bring 1 smaller than 6, which is 5.

Review

Reviewing 1 More and 1 Less

1 less

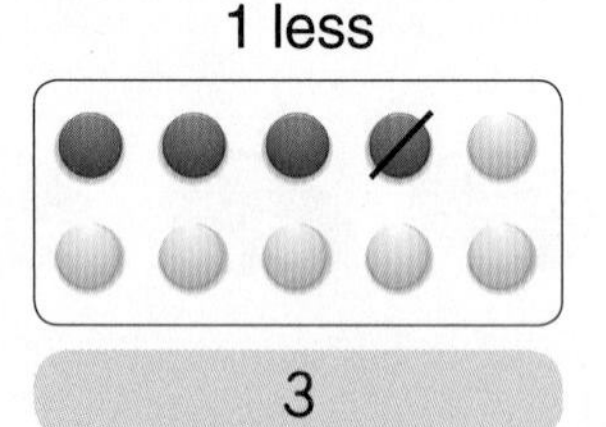

3

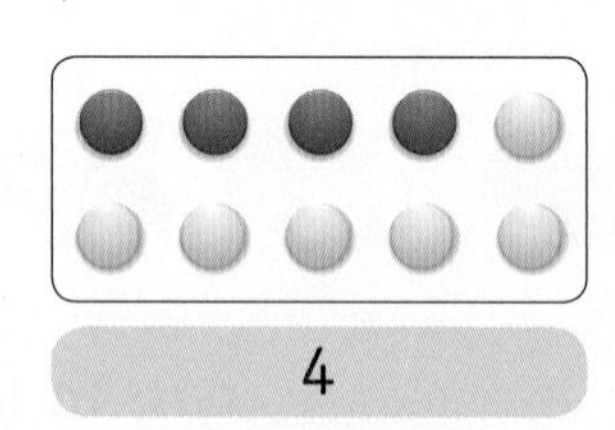

4

1 more

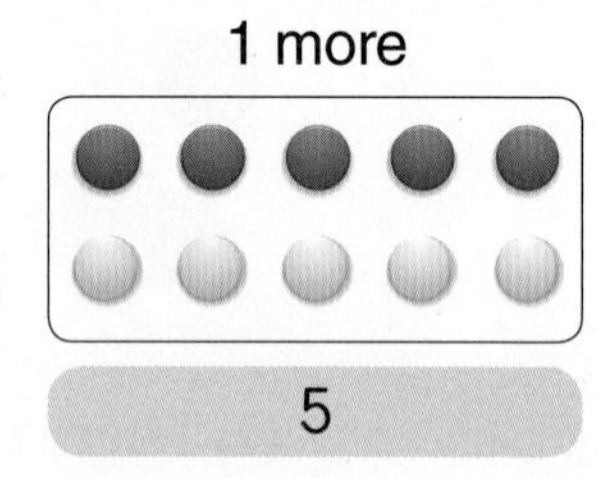

5

Number that is 1 more: 1 more than the current number. When writing the numbers in order, it comes right after the current number.

Number that is 1 less: 1 less than the current number. When writing the numbers in order, it comes right before the current number.

Warm Up 08 Draw a circle next to the correct statement.

(1)
8 is 1 more than 7. ( )
8 is 1 less than 7. ( )

(2)
5 is 1 more than 6. ( )
5 is 1 less than 6. ( )

08-1 **Read the phrase. Draw the correct number of circles in the box below.**

1 more than 7 ➡ [ ]

08-2 **Look at the picture. Write the correct number in the blanks.**

Cars Robots Teddy bears

(1) 1 less than the total number of teddy bears is ______.

(2) 1 more than the total number of robots is ______.

(3) The number of cars is ______ less than the number of robots.

Numbers Smaller than 20

Comparing Numbers up to 9

Comics for Learning Math

A little boy is crying on the street because he is hungry. Between Ryan and Aris, whoever has the greater number of snacks will give some to the boy.

Ryan has 4 cookies while Aris has 7. Therefore, Aris has more cookies.

Review

Comparing Numbers up to 9

- When you compare two numbers, you can say "bigger than" or "smaller than."
- When you compare two or more things, you can say "more than" or "fewer than."

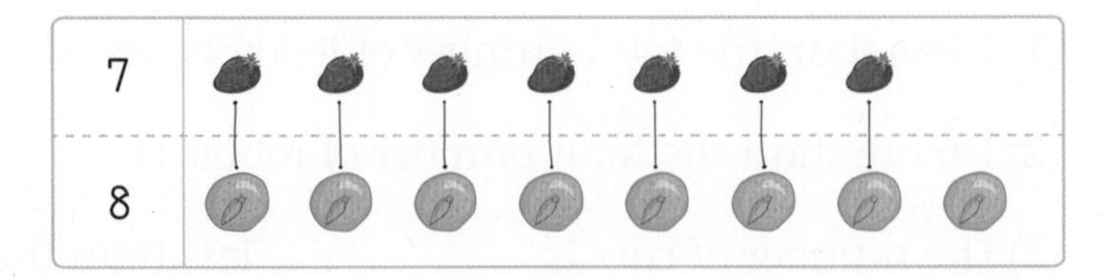

- 8 is bigger than 7. There are more oranges than strawberries.
- 7 is smaller than 8. There are fewer strawberries than oranges.

Warm Up 09 **Count the number of robot and teddy bears. Circle the correct word.**

There are (more / fewer) robots than teddy bears.

09-1 **Circle the bigger number.**

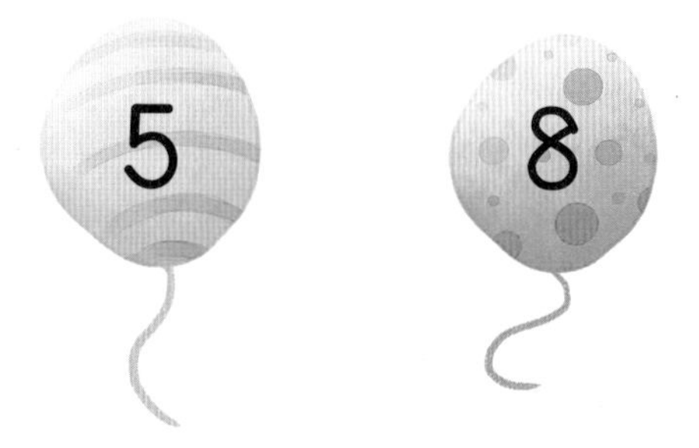

09-2 **Which sentence is incorrect?**

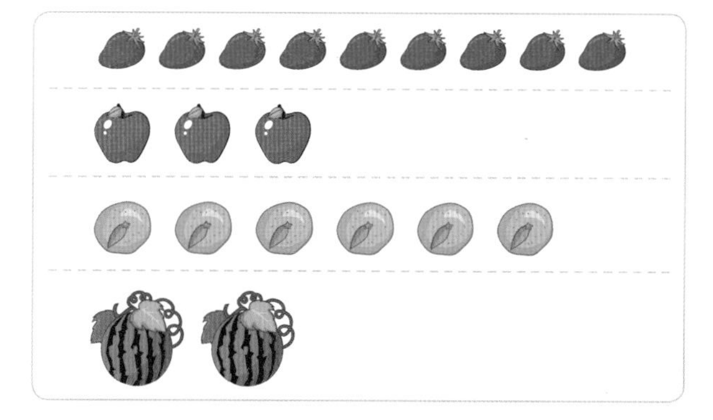

a. There are more strawberries than apples.

b. There are more apples than oranges.

c. There are fewer oranges than strawberries.

d. The number of watermelons is the smallest.

Numbers Smaller than 20

Different Ways of Grouping the Numbers 2, 3, 4, 5

Comics for Learning Math

Two factory workers are talking about how to deal with four malfunctioning robots. To stop the robots, they need to take the Stones of Life out of each robot.

Review

Separating and Adding 2

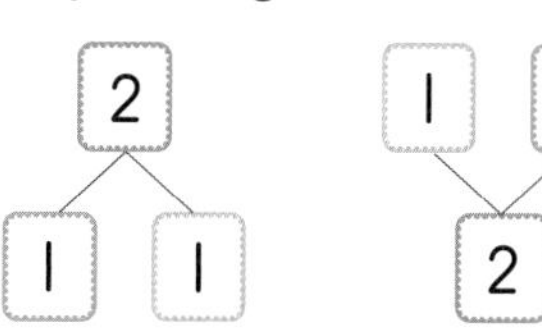

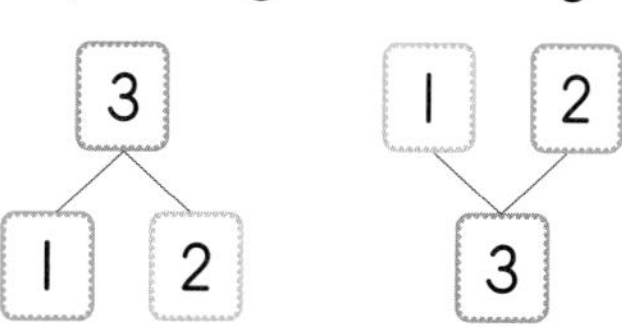

You can separate 2 into (1 and 1).
If you add together (1 and 1), you get 2.

Separating and Adding 3

You can separate 3 into (1 and 2) and (2 and 1).
If you add together (1 and 2) or (2 and 1), you get 3.

Separating and Adding 4

Example

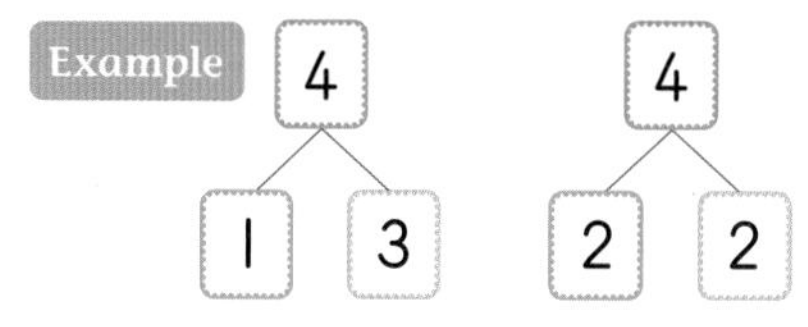

You can separate 4 into (1 and 3), (2 and 2), or (3 and 1).

Example

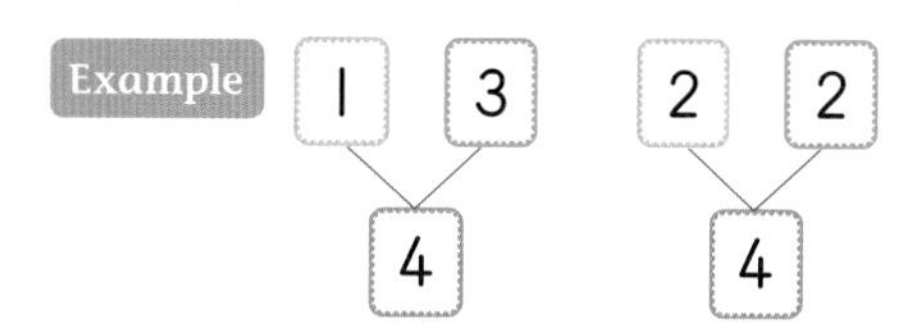

If you add together (1 and 3), (2 and 2), or (3 and 1), you get 4.

Separating and Adding 5

Example

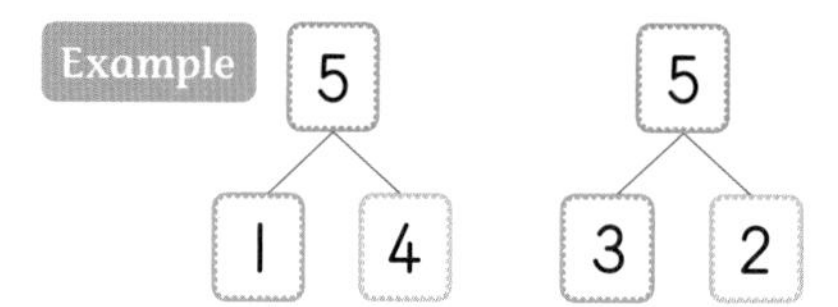

You can separate 5 into (1 and 4), (2 and 3), (3 and 2), or (4 and 1).

Example

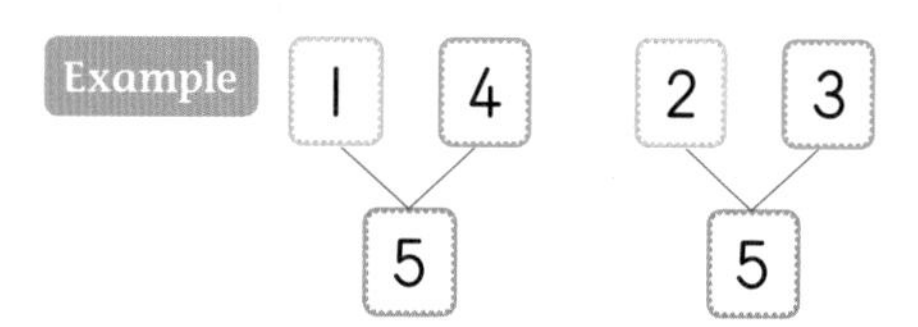

If you add together (1 and 4), (2 and 3), (3 and 2), or (4 and 1), you get 5.

Warm Up 10 **Draw the correct number of ♥ in the box.**

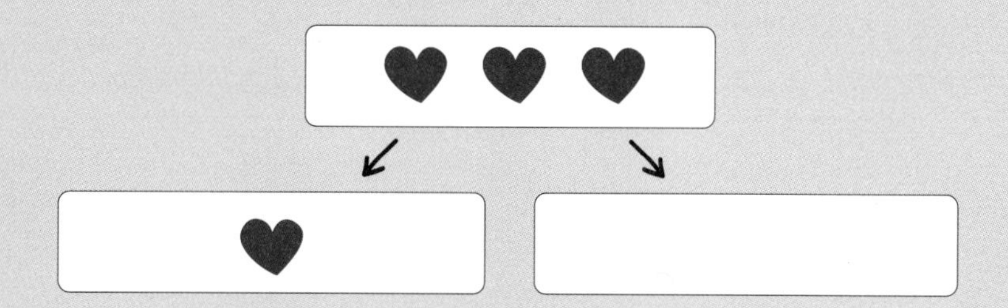

10-1 Write the correct number in the □.

10-2 Which of the below is <u>incorrectly</u> separated or added?

a. 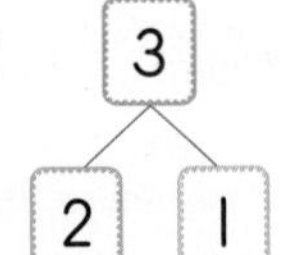

b. 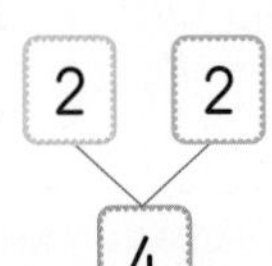

c. 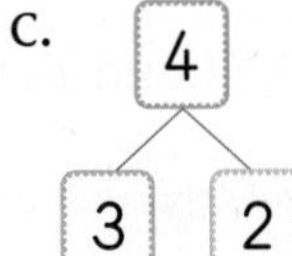

d. 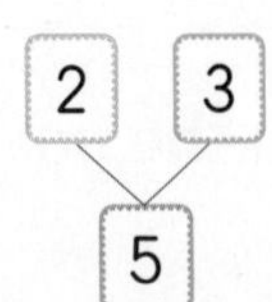

e. 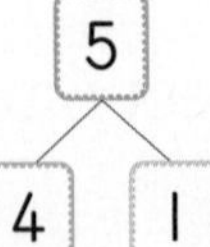

Numbers Smaller than 20

Different Ways of Grouping the Numbers 6, 7, 8, 9

Comics for Learning Math

Ryan and Aris are looking for other malfunctioning robots in the woods. They come across the same two factory workers. The workers are talking about something.

8 can be separated, or grouped, into (1, 7), (2, 6), (3, 5) and (4, 4).

Review

Separating and Grouping 6

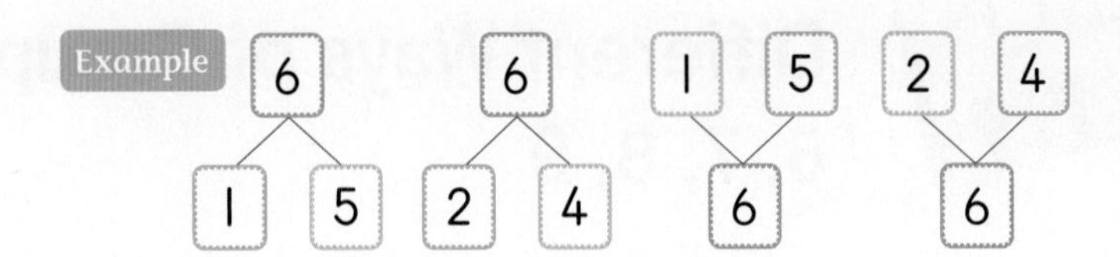

- You can separate 6 into (1 and 5), (2 and 4), (3 and 3), (4 and 2), or (5 and 1).
- If you add together (1 and 5), (2 and 4), (3 and 3), (4 and 2), or (5 and 1), you get 6.

Separating and Grouping 7

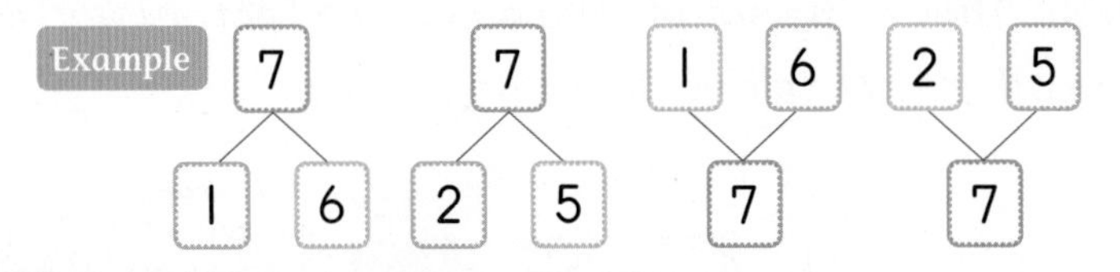

- You can separate 7 into (1 and 6), (2 and 5), (3 and 4), (4 and 3), (5 and 2), or (6 and 1).
- If you add together (1 and 6), (2 and 5), (3 and 4), (4 and 3), (5 and 2), or (6 and 1), you get 7.

Separating and Grouping 8

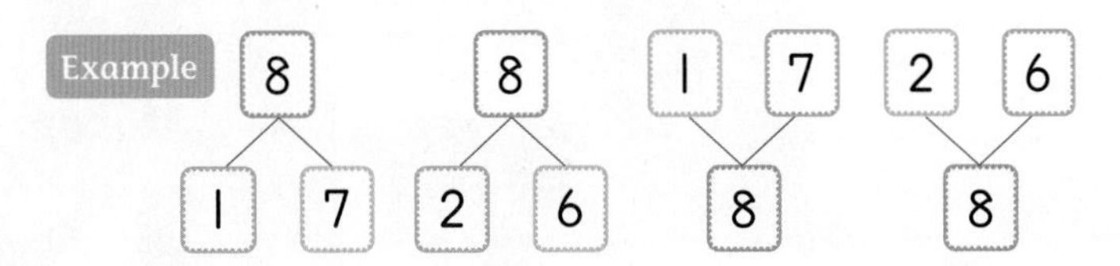

- You can separate 8 into (1 and 7), (2 and 6, (3 and 5), (4 and 4), (5 and 3), (6 and 2), or (7 and 1).
- If you add together (1 and 7), (2 and 6), (3 and 5), (4 and 4), (5 and 3), (6 and 2), or (7 and 1), you get 8.

Separating and Grouping 9

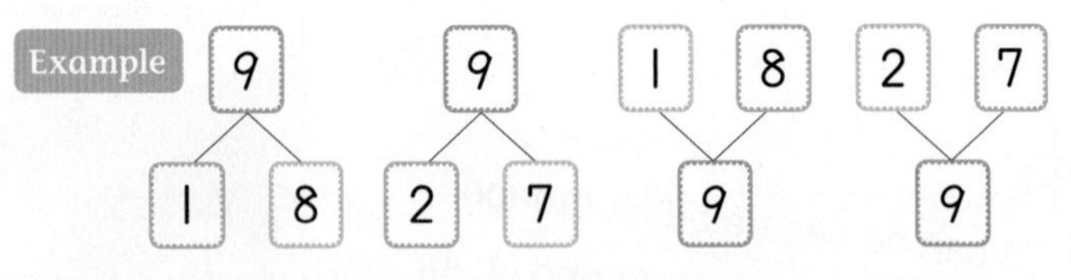

- You can separate 9 into (1 and 8), (2 and 7), (3 and 6), (4 and 5), (5 and 4), (6 and 3), (7 and 2), or (8 and 1).
- If you add together (1 and 8), (2 and 7), (3 and 6), (4 and 5), (5 and 4), (6 and 3), (7 and 2), or (8 and 1), you get 9.

Warm Up 11 Draw in the box the total number of oranges gathered.

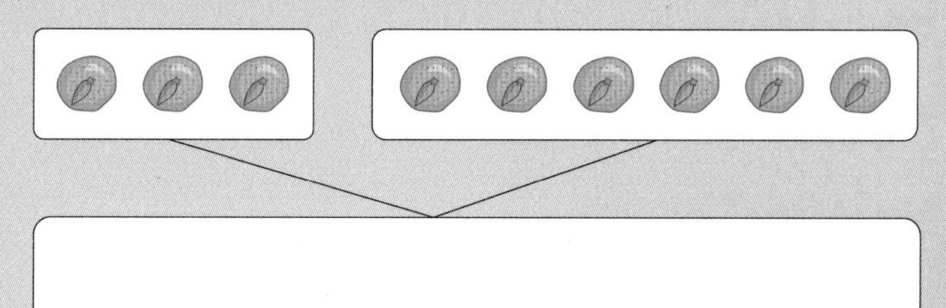

11-1 Write the correct number in the ☐.

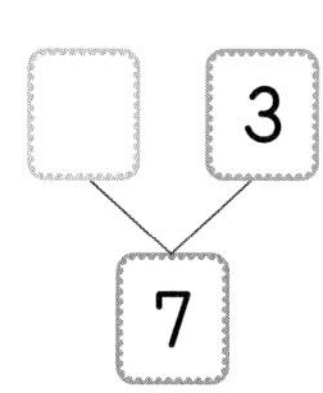

11-2 Which pair of numbers <u>cannot</u> be put into the ☐s?

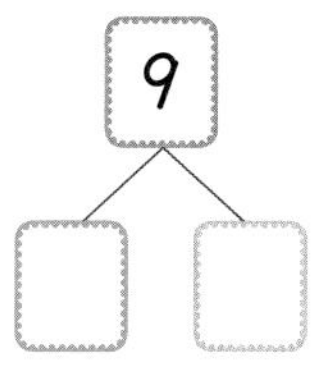

a. 4 5

b. 2 8

c. 3 6

d. 1 8

e. 2 7

Numbers Smaller than 20

Understanding, Reading, and Writing 10

Comics for Learning Math

More robots have gathered, and now there are more than before. Aris asks Ryan how many robots there are altogether.

1 greater than 9 is 10.
It is read as "ten."

Review

Understanding, Writing, and Reading 10

1 more than 9 is 10.
You read 10 as "ten."

Warm Up 12 **How many strawberries are there?**

12-1 **Choose the correct statement.**

a. 1 more than 8 is 10.
b. 10 is read as "ten."
c. 10 is read as "tenth."
d. 1 less than 9 is 10.

12-2 **Which of the sentences below is using 10 incorrectly?**

a. I have 10 colored pencils.
b. My older sister is 10.
c. At the grocery store I was 10 in line.
d. There are 10 candies on the desk.

Numbers Smaller than 20

Grouping and Gathering the Number 10

Comics for Learning Math

Ryan and Aris have gathered 10 Stones of Life.
They are trying to bring the stones to Sir Walter.

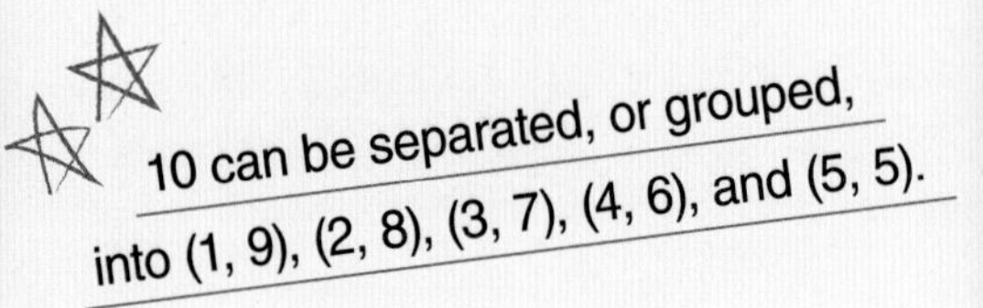

Review

- 1 more than 9.
- 2 more than 8.

- You can separate 10 into (1 and 9), (2 and 8), (3 and 7), (4 and 6), (5 and 5), (6 and 4), (7 and 3), (8 and 2), or (9 and 1).
- If you add together (1 and 9), (2 and 8), (3 and 7), (4 and 6), (5 and 5), (6 and 4), (7 and 3), (8 and 2), or (9 and 1), you get 10.

Warm Up 13

Draw the correct number of 🍭.

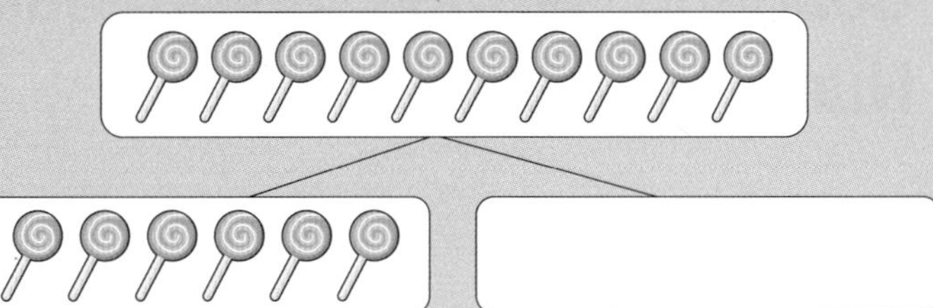

13-1

What number fits into both ☐s?

- 2 more than 8 is ☐.
- ☐ is the number you get if you group 4 and 6.

a. 2 b. 6 c. 8 d.9 e. 10

13-2

Which of the following explanations is NOT correct?

a. You can separate 10 sheep into 3 sheep and 7 sheep.

b. When you group together 9 apples and 1 apple, there will be 8 apples.

c. When you group 7 fish and 3 fish, there will be 10 fish.

d. You can separate 10 crayons into groups of 5 and 5.

e. When you group together 9 strawberries and 1 strawberry, there will be 10 strawberries.

Numbers Smaller than 20

Counting, Writing, and Reading Numbers from 11 to 19

Comics for Learning Math

Aris wants to count the number of Stones of Life that Sir Walter has taken out of the robots, but she is having a hard time counting higher than 10.

If you are expressing a number greater than 10, you write the number of tens in the tens place and the number of ones in the ones place.

Review

Write	Say
11	Eleven
12	Twelve
13	Thirteen
14	Fourteen
15	Fifteen
16	Sixteen
17	Seventeen
18	Eighteen
19	Nineteen

Warm Up 14 Count the oranges and write the number on the line.

14-1 Count the bells. Choose the correct written and spoken form of the number.

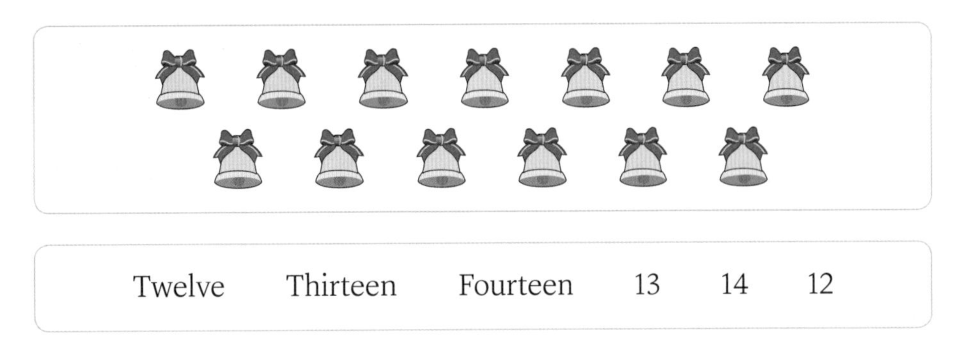

Twelve Thirteen Fourteen 13 14 12

14-2 Who is saying the number incorrectly? Write their name on the line.

Numbers Smaller than 20

Grouping and Gathering Numbers from 11 to 19

Comics for Learning Math

Aris is excited to have learned how to count numbers bigger than 10. Ryan gives Aris a problem to solve. He asks her to express the number 13 using the Stones of Life.

You can express 13 with 1 group of ten and 3 ones.

Review

Number	Tens		Ones	
11		1		1
12		1		2
13		1		3
14		1		4
15		1		5
16		1		6
17		1		7
18		1		8
19		1		9

Warm Up 15 **Look at the picture and write the correct number in the ☐.**

When a bag of 10 teddy bears and ☐ teddy bears are combined, they make 14.

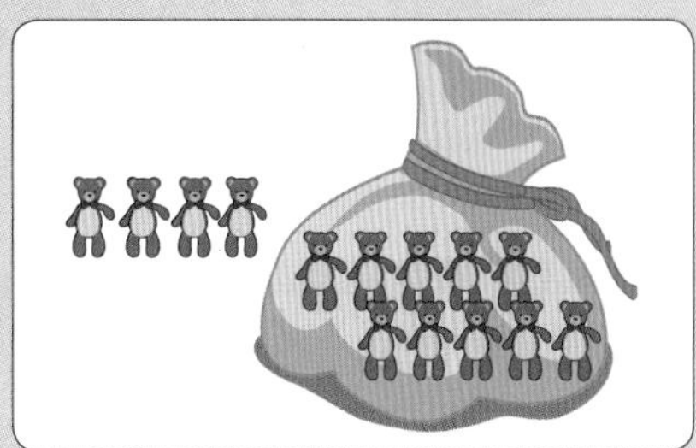

15-1 **Find the <u>two incorrect</u> descriptions of the picture.**

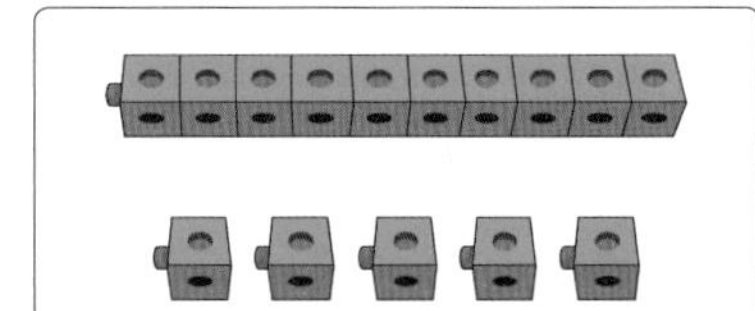

a. fifteen
b. fifteenth
c. a bundle of 10 sticks and 5 sticks
d. 10 bundles of 10 sticks and 5 sticks

15-2 **Look at the pictures. Write the correct numbers in the blanks.**

(1)

When ______ bundle of 10 sticks and ______ sticks are combined, they make ______.

(2)

You get ______ baseballs in total when you combine ______ bundle of 10 baseballs and ______ baseballs.

Numbers Smaller than 20

Understanding Numbers up to 19

Comics for Learning Math

Sir Walter has gathered parts to make a robot at home.
Aris has forgotten the number of the robot she wants.

She chooses the robot that is one after 12 and one before 14, which is 13.

Review

Understanding the Numbers up to 19

1	2	3	4	5	6	7	8	9	10
11	12	13	14	15	16	17	18	19	

➡ 13 is between 12 and 14.

Warm Up 16 What number goes into the ☐?

14 – ☐ – 16

a. 19 b. 17 c. 15 d. 13 e. 11

16-1 Write the correct numbers in each ☐.

9 – 11 – 13 – ☐ – ☐

16-2 If you put the numbers below in order from biggest to smallest, which number is the second biggest?

12, 17, 9, 11, 19, 10, 16

➡ ☐, ☐, ☐, ☐, ☐, ☐, ☐

a. 19 b. 17 c. 12 d. 10 e. 9

Numbers Smaller than 20

Comparing Numbers Up to 19

Comics for Learning Math

After getting all the parts necessary to make a robot, they are ready to go home. However, Ryan and Aris are curious about the numbers that are posted on the wall of the factory.

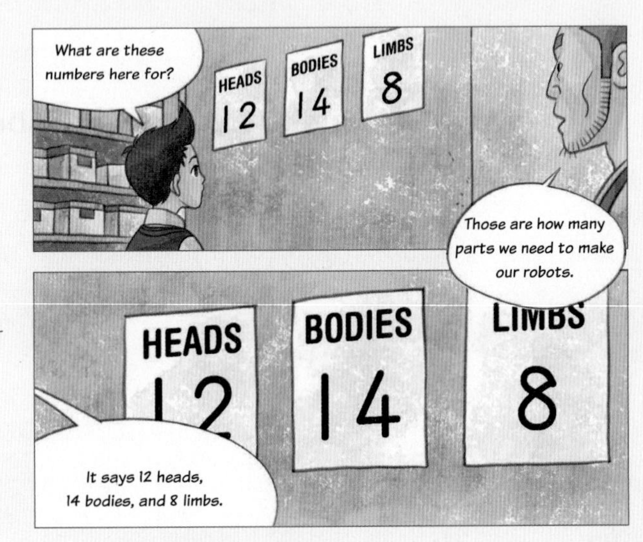

If you want to compare three numbers, you can do this easily by comparing two numbers at a time.

Review

When the numbers in the tens place are different, the side with the larger number is the greater number.

7

11

➡ $7<11$, $11>7$

Read 7 is less than 11. 11 is greater than 7.

When the numbers in the tens place are the same, then the side with the larger number in the ones is the greater number.

13

18

➡ $13<18$, $18>13$

Read 13 is less than 18. 18 is greater than 13.

13

16

14

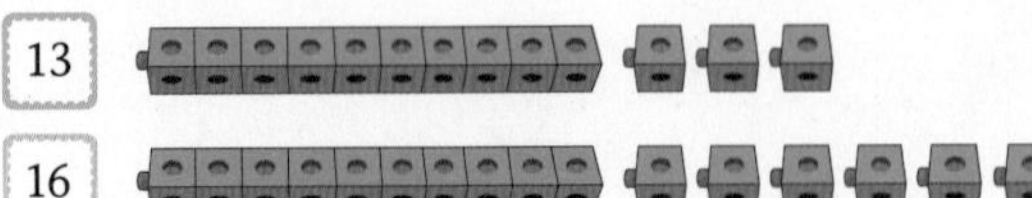

➡ $13<14<16$, $16>14>13$

Read 14 is greater than 13, and 14 is less than 16.

Warm Up 17 Look at the pictures. Which one is smaller?

17-1 **Which one is correct?**

a. $11 > 14$

b. 15 is greater than 19.

c. $17 < 16$

d. 12 is less than 16.

17-2 **Look at the comparisons below and write the correct number in the blanks.**

$15 > 12$ and $15 < 17$.

Among 15, 12, and 17, the biggest number is ______ and the smallest number is ______.

Numbers Smaller than 20

Understanding Even and Odd Numbers

Comics for Learning Math

There are boxes full of robot parts in Sir Walter's workroom. Ryan and Aris are sorting and counting each item.

If you can make a pair, that is an even number, and if you have a leftover, that is called an odd number.

Review

Even and Odd Numbers

Even numbers Numbers that can be separated into two equal groups. All even numbers have 0, 2, 4, 6, or 8 in the ones place. All even numbers can be made into pairs with no remainder.

Odd numbers Numbers that cannot be separated into two equal groups. All odd numbers have 1, 3, 5, 7, or 9 in the ones place. All odd numbers, if made into pairs, will have one remainder.

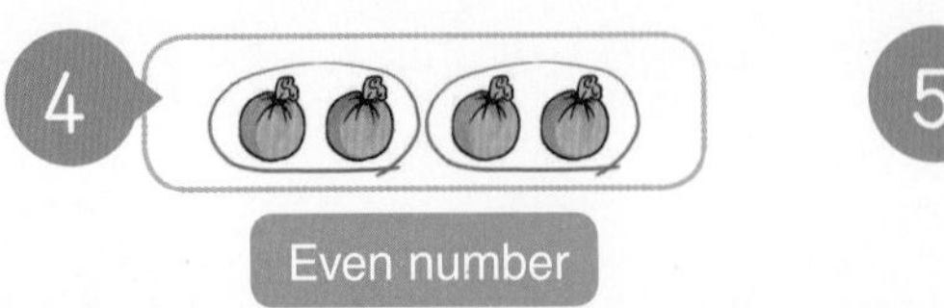

Even number

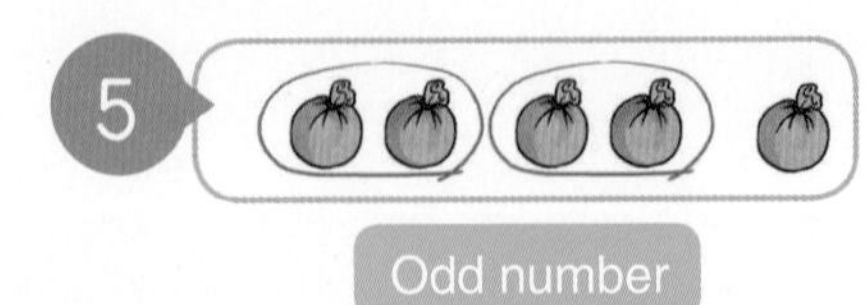

Odd number

Warm Up 18 Circle sets of two. Which one is an even number?

a. 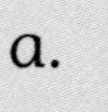 

b. 

c. 

d. 

e. 

18-1 Which series has all even numbers?

11, 14, 18, 17, 12, 15

a. 14, 18
b. 18, 12, 15
c. 11, 17, 15
d. 14, 18, 12

18-2 How many odd numbers are greater than 12 but less than 18?

Understanding the Number 10

Look at the picture. There are number cards placed in a pattern. Think carefully about the following questions and write the answers.

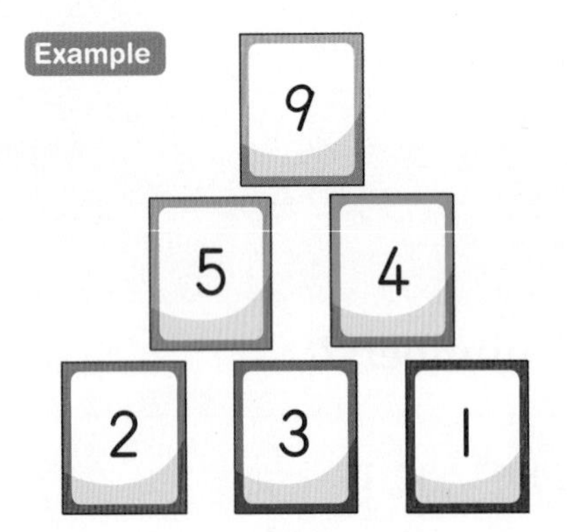

What kind of connection do 5 and 4 have with 9?
What kind of connection do 2 and 3 have with 5?

1 Let's look for the pattern in the Example. Once you find it, explain the pattern below.

2 Fill in the boxes with numbers to complete the pattern. You may use the same number more than once.

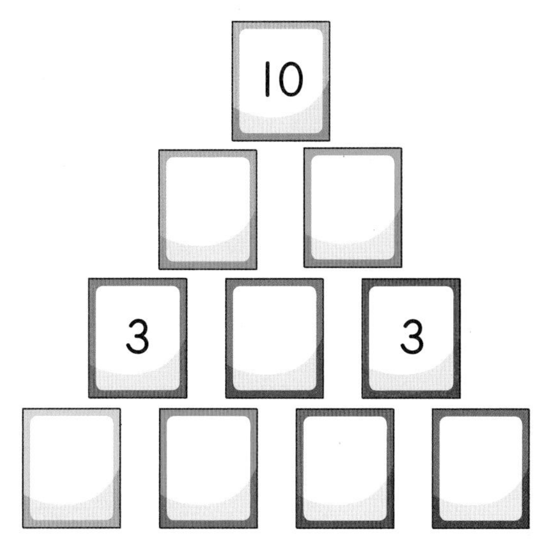

For example, you can separate 10 into 1 and 9. Does that work? Is there any other way?

After you separate 10 into 2 numbers on the second row, you can think about what will work for the third row.

Understanding Numbers up to 19

Ryan and Aris are playing with number cards. Read the dialog and answer the question.

Try to guess which number card I have. You can ask me 4 "yes or no" questions.

Okay, I'll start. Is the number greater than 10?

Is the number even?

Is the number greater than 5?

Hmm…
I have one chance left.

1 Circle all the numbers that Ryan could have.

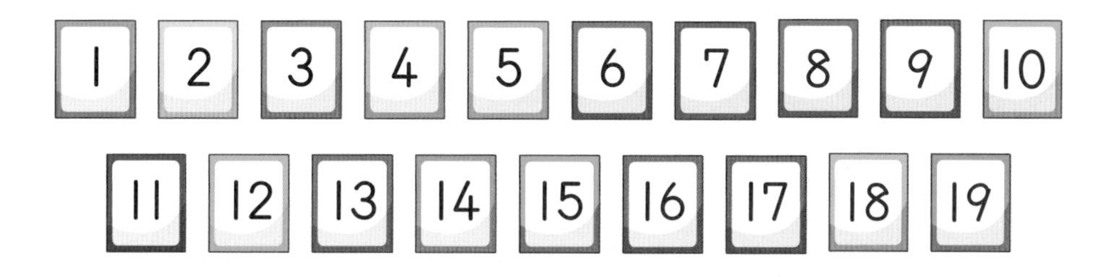

2 If you were Aris, what would you ask to find out which number card Ryan has?

You can ask only one question.
Think carefully.

Understanding Numbers up to 19

Ryan and Aris are playing a pattern recognition game. Look at the number board Ryan has and answer the question.

Hey, Aris! Did you finish filling in your board? Let's start!

Okay, you go first!

Hmm...Oh! I found a pattern on my board!

What's a "pattern"?

Look here. This line has the numbers 3, 6, 9, and 12.
The number increases by 3 each time!

Really? Let's see if we can find another pattern!

Sure!

Ryan

1	2	3	4
7	14	6	5
11	10	9	8
13	16	12	15

Write down two more patterns from Ryan's board.

① ______________________________

② ______________________________

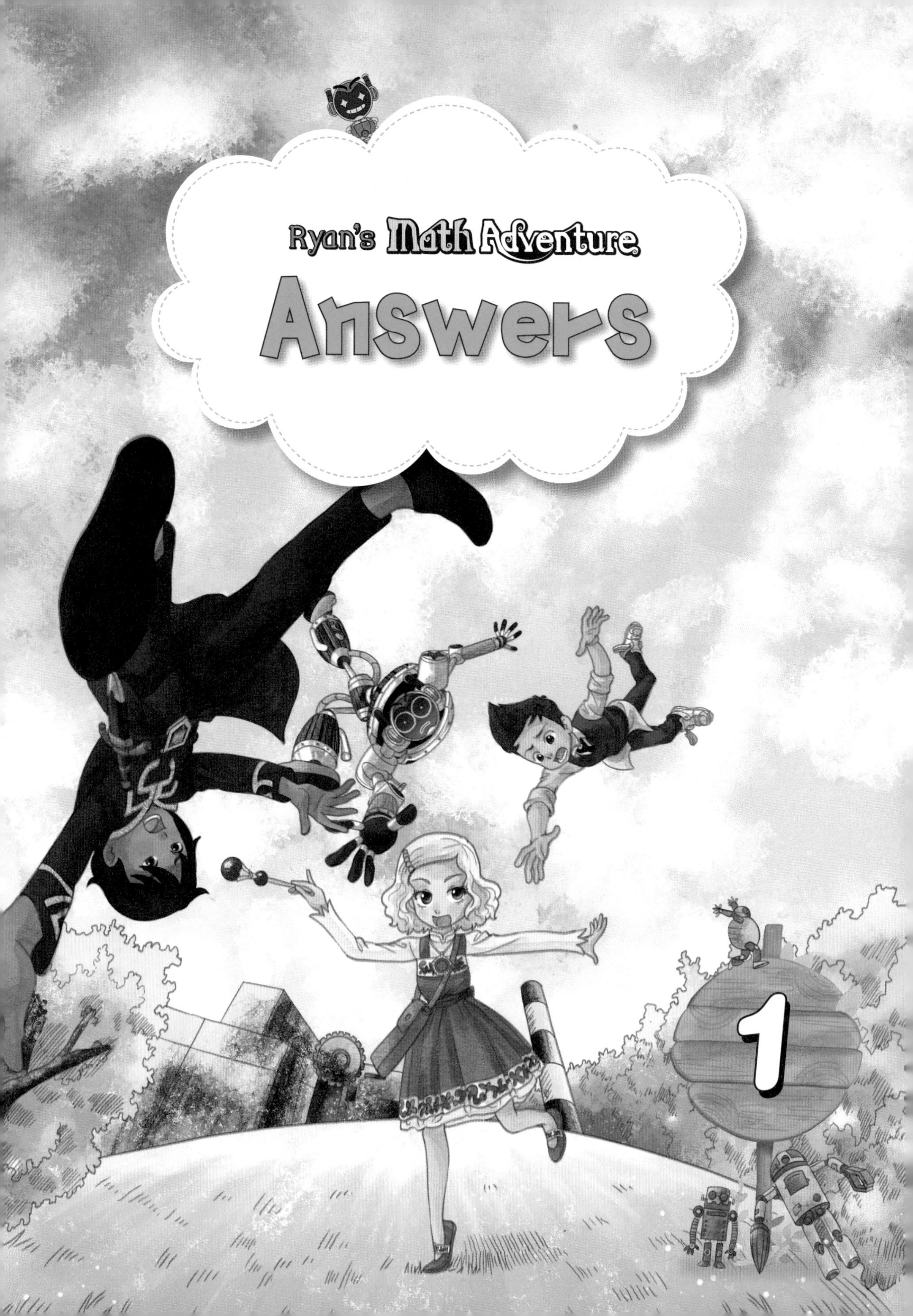
Ryan's Math Adventure
Answers
1

Answers

1 Counting, Writing, and Reading Numbers from 1 to 5

p. 133

Warm Up 01 c

01-1 b

01-2 (1) 2 (2) 3 (3) 4 (4) 5

Warm Up 01

c. There are 3 apples but 4 sticks.

01-1

b. If there are 2 strawberries, you read it as "two" and write "2."

01-2

There is one (1) paper boat, two (2) glue sticks, two (2) paper clips, three (3) pencils, four (4) pairs of scissors, five (5) erasers, and five (5) pieces of construction paper.

2 Ordering from 1 to 5

p. 135

Warm Up 02 c

02-1 d

02-2 (1) Avan (2) Mika

Warm Up 02

Alice is third.

02-1

1 is first, 2 is second, 3 is third, 4 is fourth, and 5 is fifth.

02-2

The first child in line is Avan. Mika is the third child in line.

3 Understanding the Idea of 1 More and 1 Less (1)

p. 137

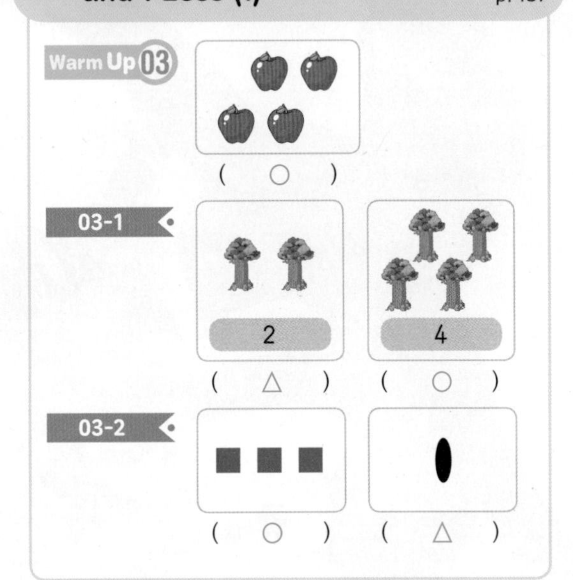

Warm Up 03

1 more than 3 is 4. Find the picture with 4 apples.

03-1

1 more than 3 is 4.

1 less than 3 is 2.

03-2

1 more than 2 is 3.

1 less than 2 is 1.

Answers

4 Understanding, Writing, and Reading 0 p. 139

Warm Up 04 0, Zero

04-1 0

04-2 (1) 0 (2) 0

Warm Up 04

There are no apples left. The number of apples is "0" and read as "zero."

04-1

There are 3, then 2, and then 1. The numbers are getting smaller by 1. The answer is 0.

04-2

(1) 1 less than 1 is "0" and read as "zero."

(2) 1 more than 0 is "1."

5 Counting, Writing, and Reading Numbers from 6 to 9 p. 141

Warm Up 05 6

05-1

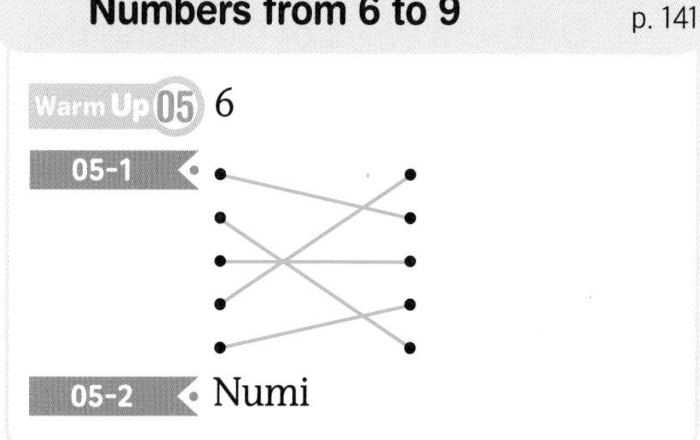

05-2 Numi

Warm Up 05

1 more than 5 is written as "6" and read as "six."

05-2

There are 6 hearts, which is one more than 5. You can read it as "six."

6 Understanding the Order of Numbers up to 9 p. 143

Warm Up 06 4th, 8th

06-1 b

06-2 (1) Sixth (2) Eighth

Warm Up 06

The number that comes after 3 is 4. The number after 7 is 8.

06-1

b. 4-5-6-7-8 can be read as Fourth-Fifth-Sixth-Seventh-Eighth when you read them as ordinal numbers.

06-2

(1) The orange is the sixth fruit in the row.

(2) The kiwi is the eighth fruit in the row.

7 Comparing Numbers p. 145

Warm Up 07 Aris

07-1

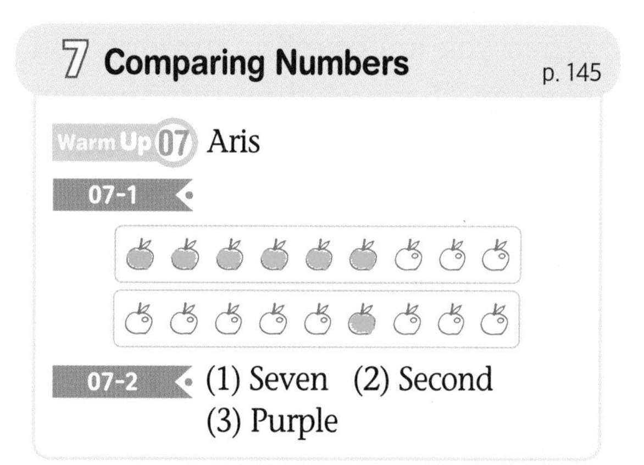

07-2 (1) Seven (2) Second (3) Purple

Answers

Warm Up 07

Count the strawberries. One, two, three, four, five. There are five strawberries. Fifth is an ordinal number.

07-1

When you see a number like six, you need to color six objects. When you see a number like sixth, you need to color only the sixth object in the line.

8 Understanding the Idea of 1 More and 1 Less (2)

p. 147

Warm Up 08

(1) 8 is 1 more than 7. (○)

(2) 5 is 1 less than 6. (○)

08-1 ○○○○○○○○

08-2 (1) 7 (2) 8 (3) 1

Warm Up 08

When you are counting upward, you can say "1 more" than the number. When you are counting downward, you can say "1 less" than the number.

08-1

1 more than 7 is 8.

08-2

There are 6 cars, 7 robots, and 8 teddy bears.

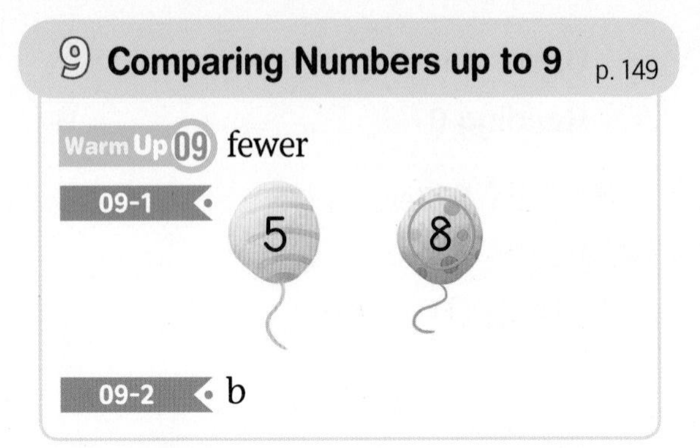

9 Comparing Numbers up to 9

p. 149

Warm Up 09 fewer

09-1

09-2 b

Warm Up 09

There are 4 robots and 7 teddy bears. The number of robots is smaller than the number of teddy bears.

09-1

8 is greater than 5.

09-2

b. There are 3 apples and 6 oranges. The number of apples is smaller than oranges.

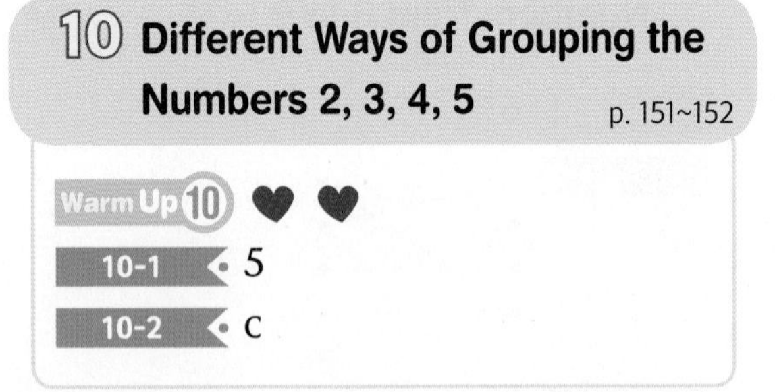

10 Different Ways of Grouping the Numbers 2, 3, 4, 5

p. 151~152

Warm Up 10 ♥ ♥

10-1 5

10-2 c

Warm Up 10

You can separate 3 into 1 and 2.

10-1

When you group 4 and 1, you'll get 5.

Answers

10-2
c. 4 can be separated into 3 and 1 or 2 and 2.

11 Different Ways of Grouping the Numbers 6, 7, 8, 9 p. 155

Warm Up 11 ○○○○○○○○○
11-1 4
11-2 b

Warm Up 11
When you put 3 oranges and 6 oranges together, you have 9 oranges.

11-1
When you put 4 and 3 together, you have 7.

11-2
b. You can separate 9 into 2 and 7.

12 Understanding, Reading, and Writing 10 p. 157

Warm Up 12 10
12-1 b
12-2 c

Warm Up 12
There are 10 strawberries in total.

12-1
b. You read 10 as "Ten."

12-2
c. When talking about your place in line, you must use the ordinal "tenth."

13 Grouping and Gathering the Number 10 p. 159

Warm Up 13 ○○○○
13-1 e
13-2 b

Warm Up 13
You can separate 10 into 6 and 4.

13-1
e. 2 is greater than 8. Putting 4 and 6 together makes 10.

13-2
b. When you put 9 and 1 together, you have 10.

14 Counting, Writing, and Reading Numbers from 11 to 19 p. 161

Warm Up 14 11
14-1 Thirteen, 13
14-2 Ryan

Warm Up 14
There are 11 oranges in total.

14-1
You read 13 as "Thirteen."

Answers

14-2

Ryan : You have to read the date as "Today is the fifteenth."

15 Grouping and Gathering Numbers from 11 to 19 p. 163

Warm Up 15 4

15-1 b, d

15-2 (1) 1, 8, 18 (2) 13, 1, 3

Warm Up 15

14 is a grouping of 10 teddy bears with 4 teddy bears.

15-1

15 is read as "fifteen." It is a grouping of 10 and 5.

16 Understanding Numbers up to 19 p. 165

Warm Up 16 c

16-1 15, 17

16-2 b

Warm Up 16

1 greater than 14 is 15. 15 comes after 14. 16 comes after 15.

16-1

11 comes after 9. 13 comes after 11. The patter is that the next number is 2 greater than the previous number. Therefore, 15 is after 13. 17 is after 15.

16-2

If you arrange the numbers from greatest to smallest, the greatest number is 19. 17 is the second greatest number.

17 Comparing Numbers up to 19 p. 167

Warm Up 17 13

17-1 d

17-2 17, 12

Warm Up 17

1 ten and 3 ones is 13. 1 ten and 7 ones is 17.

The smaller number is the number has less ones, which is 13.

17-1

d. 12 is less than 16. You can express the relationship like this, "12<16."

17-2

15 is greater than 12, but 15 is less than 17.

Answers

18 Understanding Even and Odd Numbers

p. 169

Warm Up 18 c
18-1 d
18-2 3

Warm Up 18

c. You can call this type of number an "even number."

18-1

d. You can make pairs with 14, 18, and 12. There will be remainders with 11, 17, an 15.

18-2

The odd numbers are 13, 15, and 17. There are 3 odd numbers.

Understanding the Number 10

p. 170~171

1 As you go downward, you separate the number on top.
2 5, 5, 2, 2, 1, 1, 2

1 As you go downward, you separate the number on top.

Reason

You can separate 9 into 5 and 4, 5 into 2 and 3, and 4 into 3 and 1

2 You can separate 10 into (1, 9), (2, 8), (3, 7), (4, 6), (5, 5), (6, 4), (7, 3), (8, 2), and (9, 1). The pairs (1, 9), (2, 8), (3, 7), (7, 3), (8, 2) and (9, 1) either have the number 3 or a number smaller than 3. Therefore, you can only separate 10 into (4, 6), (5, 5), and (6, 4). Let's think about which number cards are needed for the next line.

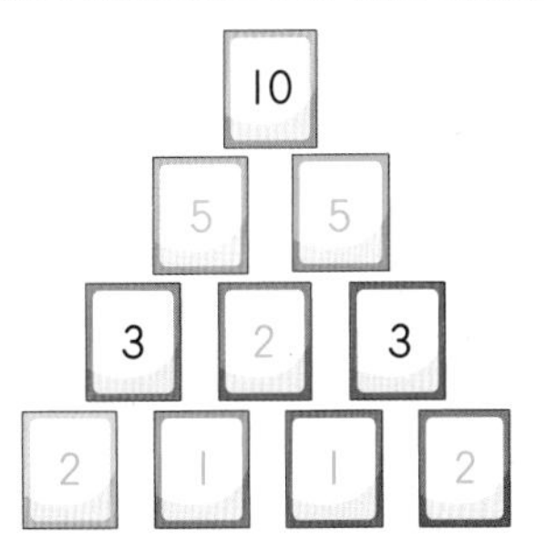

Answers

Understanding Numbers up to 19

p. 172~173

1 7 9

2 Example Is the number 7?

1 Aris asked if the number is greater than 10. Ryan said, "No." We can erase all the numbers that are greater than 10. Only 1, 2, 3, 4, 5, 6, 7, 8, 9, and 10 are left. Aris next asked if the number is even. Ryan said, "No." We can only choose the odd numbers 1, 3, 5, 7, and 9. Aris then asked if the number is greater than 5. Ryan said, "Yes." Only 7 and 9 are left.

2 With one more question, you can find out if the number card Ryan has is either 7 or 9.

Understanding Numbers up to 19

p. 174

1 Example The horizontal line of 1, 2, 3, and 4 is written in numerical order. (They increase by 1 each time.)

2 Example The horizontal line of 11, 10, 9, 8 is written in reverse numerical order. (They decrease each time.)

MEMO

MEMO

Ryan's Math Adventure

Book	Subject	Area	Grade
1	Numbers Smaller than 20	Numbers and Calculation	CK, C1
2	Basics of Addition and Subtraction	Numbers and Calculation	CK, C1, C2
3	Basics of Figures	Figures	CK, CK-2, C1, C2, C4, SK-2
4	Decimal System	Numbers and Calculation	C1, C2, C3
5	Addition and Subtraction	Numbers and Calculation	C1, C2, C3
6	Basics of Measurement	Measurement	CK
7	Arithmetic Sense (1)	Numbers and Calculation	C1, C2, C3
8	Time	Measurement	C1, SK-2, C2, C3, C4
9	Basics of Multiplication	Numbers and Calculation	C2, SK-2, S3-5, C3
10	Spatial Sense	Figures	CK, SK-2
11	Length	Measurement	C2, S3-5, SK-2
12	Movement of Figures	Figures	SK-2
13	Basics of Division	Numbers and Calculation	C3
14	Basics of Data Collection and Expression	Data and Probabilities	CK, C1, SK-2, C2, C3
15	Capacity	Measurement	C3, C4
16	Meaning of Fractions	Numbers and Calculation	C1, C2, C3, SK-2
17	Weight	Measurement	C3, C4, S3-5
18	Multiplication and Division	Numbers and Calculation	C3, C4, S3-5
19	Meaning of Decimals	Numbers and Calculation	C4, S3-5, C5
20	Various Rules	Regularity	SK-2, C3
21	Big Numbers and Numeric Sense	Numbers and Calculation, Integrated Measurement	C3, C4, C5
22	Angles and Triangles	Figures, Integrated Measurement	C4, C5
23	Arithmetic Sense (2)	Numbers and Calculation	C3, C4, C5, S3-5, C6
24	Polygon (Rectangle)	Figures	C4, C5
25	Regularity and Generalization	Regularity	C4, C5, C6
26	Divisors and Multiples	Numbers and Calculation	C2, C4, C6
27	Congruence, Symmetry	Figures	C4, S3-5, C7, C8
28	Circumference and Area	Measurement	C3, C4, C5, S6-8
29	Adding and Subtracting Fractions	Numbers and Calculation	C3, C4, C5
30	Probability and Representative Value	Data and Probabilities	SK-2, C6, S3-5
31	Proportion and Ratio	Regularity	C6
32	Circles, Circular constant, and Circular Area	Figures and Integrated Measurement	C4, S6-8, C7
33	Multiplication and Division of Fractions	Numbers and Calculation	C4, C5, C6
34	Solid Figures	Figures	CK, C1, C2, S3-5, C6, C7
35	Arithmetic Calculation of Decimals	Numbers and Calculation	C4, C5, S3-5
36	Dividing Decimals	Numbers and Calculation, Integrated Regularity	C4, S3-5, C5, C6, S6-8
37	Surface Area and Volume	Measurement	C5, C6, S3-5, S9-12
38	Tables and Graphs	Data and Probabilities	C3, C5, S3-5
39	Proportional Expression: Direct and Inverse Proportion	Regularity	C6, C7, S6-8
40	Numbers and Expressions	Numbers and Calculation	S3-5, C6, C7

· National Council of Teachers of Mathematics
<Principles and Standards for School Mathematics>

· National Governors Association Center for Best Practice, Council of Chief State School Officers
<Common Core State Standards for Mathematics>

"Ryan's Math Adventure effortlessly immerses students, especially students who find math unappealing, in various math concepts by portraying students with a real world use for math. This series also provides appropriate scaffold for 1st graders who already enjoy learning math but want to dig deeper into the reasoning and logic behind simple math equations and symbols. Vivid illustration and child-friendly math language not only keeps children interested, but also facilitates independent learning of math concepts that might be challenging on their own."

-Nina Lim (1st grade teacher at Stratford School, California)

"Ryan's Math Adventure allows students to apply mathematical concepts to the world around them. The book is distinct in that it moves away from the typical formula of rote memorization and instead sparks wonder and curiosity through its narrative composition."

-Alexandra Economos (3rd grade teacher at Tusculum Elementary in Nashville, Tennessee)

Ryan, Aris, and Sir Walter are searching for the prophesied Book of Light. Without math, the nearby town has been thrown into chaos. Suddenly, monstrous larvae creatures appear before Ryan and Aris! Can Ryan, Aris, and Sir Walter find the Book of Light in time to stop the evil Pesia?

$ 17.99

See you in the next book!

ISBN 979-11-87714-03-3

Shrink!
u gave me a
art attack!
WHEE
What's wrong,
Gilly?
Ryan's
Math Adventure
Onix Palace
Basics of
Addition and Subtraction
Maozha
2
PAT
Come on, you bugs!
You're no match for me!
Gilly,
that was ama

Research and Writing Staff

Chief Researcher: Jung-Sook Bang
- PhD, Louisiana State University
- Professor of elementary math education at Korea National University

Translator: Ileana Christine Quintano
- Masters of Theological Studies, Harvard University
- Bachelor of Arts in Religion, Swarthmore College

Chief Editor: Patrick Chung
- Bachelor of Arts in Economics, University of California, San Diego
- Single Subject Teaching Credential in Mathematics, San Diego State University
- Middle school math teacher at Covenant Global Christian School

RYAN'S MATH ADVENTURE 2

Published by WeDu Communications, Inc.

For information regarding permission,
write to WeDu Communications, Inc.
ADD. 18430 San Jose Ave - Unit C
City of Industry, California 91748
E-mail: info@wedugroup.com